Relations

by

Helier J. Robinson

Relations

An Improved Foundation of Mathematics,
Science, and Philosophy.

Second Edition

by

Helier J. Robinson

Sharebooks Publishing
www.sharebooks.com

Cover design: Oberon Robinson.
Back cover photo: Zenon Yaryemovitch.

Contents.

"If you were to write a history of the problems that philosophers have had with relations, you would finish up writing a complete history of philosophy." — the late Professor David Savan, University of Toronto, who was my best philosophy teacher.

Preface.

This book is written for scientists, mathematicians, philosophers, and anyone interested in these subjects. Because not all such people will be expert in all of these fields, some parts of the book will be elementary for some of the readers; I ask your indulgence for this.

The foundation of this book is the claim that relations make a better foundation for mathematics and science than does the traditional set-theoretic approach. By way of temptation to read further I mention that this foundation leads to possible new solutions to the problem of the effectiveness of mathematics in the sciences, to the problem of cosmic coincidences, to the mind-body problem, and to the problem of being.

Some readers may be outraged at some of the claims that I make. I can only assure them that they will find it quite worthwhile if they adopt a willing suspension of disbelief while continuing to read.

The book was originally intended to be a second edition of my "Relation Philosophy of Mathematics, Science, and Mind" but in the four year preparation of it there was so much new development that it is now better seen as a new book; hence it replaces the other, which, before being discontinued, was downloaded over 11,000 times.

Chapters 2, 3, 4, 5, and 9 are technical. They may be skipped on a first reading and perhaps avoided altogether by the non-technical reader.

Throughout the book words are shown in **boldface** type when defined, and are entered in the Glossary. Cross-references are shown as page numbers in parentheses; forward references usually should be ignored on a first reading. In the electronic edition of this book these cross-references are hyperlinks.

Some of the topics in the book are treated more extensively in an earlier work, *Renascent Rationalism*, published by MacMillan of Canada, Toronto, 1975 (5th. ed., Sharebooks Publishing, www.sharebooks.com, 2008), to which those readers wanting greater detail are referred. This applies particularly to Chapter 11 and later chapters.

Three perspicacious old friends of mine have my gratitude for many hours of discussion of my ideas: they are Graeme Nicholson, Emeritus Professor in the department of Philosophy, University of Toronto; Michael Pollanen, Professor of Pathology, University of Toronto, and Chief Forensic Pathologist of Ontario; and Al Vilcius, retired mathematician. I am also very grateful to my son Oberon Robinson for more discussion and for expert computer work in aid of my philosophy.

Introduction.

The thesis of this book is that relations make a better foundation for mathematics, logic, and science than do sets.

Since the time of Georg Cantor (1845-1918) and Gottlob Frege (1848-1925), followed by Russell (1872-1970), Whitehead (1861-1947), and others, the foundation of mathematics has been sets; and in parallel with this development, following George Boole (1815-1844) and what is now called Boolean algebra, modern symbolic logic was created. Each of these was praiseworthy, and important in the history of ideas, but, I here claim, flawed. For now I will deal with the flaws of basing mathematics on set theory, and how they may be corrected; for this we must look into the nature of sets and set theory.

A **set** — sometimes called a class, hence our word classification — is primitive in set theory but usually characterised as being a collection or aggregate. There are two other primitive concepts in set theory: the concepts of **member** and of **set-membership**. With these, two other concepts are defined: the **extension of a set** is the totality of its members, and the **intension of a set** is that property or properties possessed by all and only the members of the set; which is to say that every member of the set has the intension as a property, and anything that has the intension as a property is a member of the set. For example, the set of even numbers has the intension 'number divisible by two' and every number divisible by two, and only such a number, is a member of this set. A set may be identified by enumeration of its members or by stating its intension, but some sets are too large to enumerate — the set of all numbers, for example[1] — and some sets do not have intensions — such as the three-membered set

---

[1] This set is usually defined as {1,2,3...}, where the ellipsis, '...', is read as 'and so on'. This is supposed to be an enumeration of the set, but it is not: the 'and so on' contains a hidden intension, 'successor of' which defines all the numbers in the set except the first, which is enumerated.

consisting of an apple, a ten dollar bill, and the number eight. Sets which both do not have intensions and are too large to enumerate are in a kind of existential limbo: they can be named, and we may know some of their members, but we cannot know all of their members so we cannot know if such sets exist; they thus have nominal meaning only, they are nominal sets.

We next have to introduce some symbols. I apologise for this to those readers who suffered from bad math teachers and have disliked math ever since, but I assure them that the symbols do improve communication of abstract concepts such as set, member, set-membership, etc. (If this does not work then just skip all the symbolic expressions and get the sense of them from the text.) A set is represented by any upper case italic letter, such as A, B, C; set-membership is represented by ϵ — '$x \epsilon S$' means that x is a member of the set S; an enumerated set is represented by enclosing the enumeration in braces, as in {a,b,c}; and a set defined by an intension is represented by {x: xP}, where 'x:' (note the colon) is a variable — it stands for anything whatever, and reads "every x such that" — and P is the intension, so that {x: xP} reads 'the set of every x such that x is a P'. For example, the set of all natural numbers less than 10 is symbolised by {x: (x∈ℕ)∧(x<10)}, where ℕ is the set of natural numbers, the symbol '∧' stands for 'and', and the braces again indicate a set, this one defined by the intension '(x∈ℕ) ∧ (x<10)', which reads 'x is a number and is less than ten'. The enumeration of this set is given in {1, 2, 3, 4, 5, 6, 7, 8, 9}.

A special kind of set is an **ordered set**. A set is ordered if its members have to be enumerated in a particular order. Ordered sets are symbolised by angle brackets, such as ⟨a,b,c⟩. If you change the order in an ordered set then you have a different ordered set, unlike ordinary sets in which the order of the enumeration is immaterial. Thus {a,b,c}={b,a,c} but ⟨a,b,c⟩ ≠ ⟨b,a,c⟩.

Introductiion

We can now define, set-theoretically, two concepts that are fundamental in mathematics: they are *number* and *relation*. Without these concepts there would be no mathematics, so they may truly be said to be of fundamental importance in mathematics and hence science, so that to reach their definitions was a triumph for their inventors.

First, two sets are in said to be in **one-to-one correspondence** if, for each member of one set there corresponds one, and only one, member in the other set; and *vice versa*. If two sets are in one-to-one correspondence they are then said to have an equal number of members. A particular number is next defined by means of a particular set (which is an instance of that number) and the set of all sets which are in one-to-one correspondence with it; this latter is the definition of that number. For example, if I hold up my hands and define the set of them, by enumeration, then the number two is the set of every set that is in one-to-one correspondence with the set of my hands; and the set of both of my hands is an instance of the number two. In principle all the natural numbers may be defined in this way; numbers so defined are here called **extensional numbers**.

Second, the **Cartesian product** of the sets A and B, written AxB, is defined as Ax$B=\{\langle x,y \rangle: (x \in A) \wedge (y \in B)\}$. This may be read as the set of every ordered pair such that the first member of the pair is a member of A and the second member of the pair is a member of B.

An **extensional relation** between A and B is then defined as any member of the Cartesian product AxB; it is an ordered pair, a dyadic relation, a relation having two terms. Relations having more than two terms may be defined by larger Cartesians products. For example, a relation having three terms is defined as a member of AxBx$C=\{\langle x,y,z, \rangle: (x \in A) \wedge (y \in B) \wedge (z \in C)\}$. So this definition of relation presupposes the definition of number, since the size of a Cartesian product is numerical. Relations defined in this way are, in this book, called extensional relations because they are defined by means of extensions; they are logical constructs, constructed out of their terms.

Now to discuss the flaws. First, consider the number two: the set of all sets that are in one-to-one correspondence with a particular pair, where the particular pair is an instance of the number two. If the number of numbers is infinite then so is the number of pairs of numbers; so the number two is defined by means of an infinite set. Do we really want the number two to be defined *via* the concept of infinity? A small child understands the number two without any understanding of infinity. Secondly, the set of all pairs cannot be enumerated, so the definition can only work if the number two is the *intension* of the set of all pairs, rather than the *extension*; but following Frege, set-theorists have generally avoided intensions wherever possible. However, there is another flaw: set-membership and one-to-one correspondence are relations, and they are presupposed in the definition of instance of number, hence in the definition of relation; and the order of an ordered pair is also a relation and so presupposed. So the definitions of *instance of number* and of *relation* commit the fallacy of begging the question: they are circular and so vacuous. Hence if there are no alternative definitions of instance of number and of relation in set theory then these concepts have to be primitive. Thus there are three proposed corrections to set theory: instances of numbers are primitive, numbers (as opposed to their instances) are defined intensionally, and the concept of relation is primitive. This is discussed in greater detail in Section 17, Chapter 1, after we have gone into greater detail about relations.

So suppose that we make relations primitive (Chapter 1), and try to derive logic, set theory, and arithmetic (Chapters 2, 3, and 4) from them. Quite surprisingly, a most interesting chain of consequences then follows. Making relations basic leads to *mathematics*, for it turns out that our best language of relations is mathematics. This leads in turn to *theoretical*

science — particularly astronomy, physics, chemistry, and
cosmology, all of which are mathematical. Theoretical science
leads to *explanation*: if you ask a theorist what it is that
theoretical science describes, the usual answer is that
theoretical science *explains* empirical phenomena by
describing their *underlying causes*. **Explanation** is *causal*: to
describe causes is to explain their effects — and **causation** is a
relation. But why *underlying* causes? Causations are
underlying because all that we know of them is by analogy to
logical necessity, which may be characterised as singular
possibility (7, 42, 43) (as opposed to contingency, which is
plural possibility, and impossibility, which is zero possibility).
We only know causations by analogy because they are
imperceptible. As David Hume (1711-1776) pointed out, quite
correctly, we can perceive correlations, but we cannot perceive
necessities between their correlates because we never perceive
empirical necessities; and it hardly needs to be said that no
correlations — not even 100% correlations — are necessarily
causations (although they are often treated as such). So what
else is underlying, besides causal necessities? It turns out that,
provided that one is meticulous in distinguishing between
theoretical entities and the empirical evidence for them, *all*
theoretical entities are underlying. That is what '**theoretical**'
means: non-empirical. Physicists talk of seeing things like
neutrinos and quarks, but they cannot be seen — only evidence
for them may be visible. 'Underlying' means *non-empirical*,
bearing in mind that the empirical is everything known
through the senses. If we define the **empirical world** as
everything empirical, and the **noumenal world** as everything
that both exists and is non-empirical, then by definition these
two worlds are disjoint: they do not intersect, nothing
noumenal, or underlying, is empirical, and *vice versa*. (We
shall see later that this is not quite correct; but it is much
easier, this early in the book, both for ease of exposition and
for ease of understanding, to ignore this point.) But the
noumenal causally produces the empirical, and to describe

these causes is to explain the empirical. And to come full circle, the noumenal world consists of relations because theoretical science is mathematical and mathematics is our best language of relations. It turns out that the noumenal world consists of nothing but relations, some of which cause images of themselves in the empirical world, which images are empirical relations.

(Something in the noumenal world is called a **noumenon**; and something in the empirical world was called a **phenomenon** because in the past the empirical world was known as the phenomenal world. 'Empirical' is here preferred over 'phenomenal' because of being less ambiguous.)

In order to make clear that the noumenal world consists of relations — something that many readers may initially doubt — consider that all that we know scientifically of the noumenal world is relations. Lengths and durations are respectively spatial and temporal relations; relations between these are relations such as velocities, accelerations, and spatio-temporal curvature, or mass; and relations between these are relations such as momentum, energy, and force. And all of these relations are both mathematical and scientific entities.

Another reason for claiming that all noumena are relations comes from combining the discoveries of different branches of science. First note that a **structure** is a relation uniting a set of parts, and that these parts may themselves be structures. Physicists tell us that neutrons and protons are structures of quarks, that an atomic nucleus is a structure of neutrons and protons, and that an atom is a structure of a nucleus and electrons; chemists tell us that molecules are structures of atoms; microbiologists tell us that single-celled organisms are structures of molecules; biologists tell us that multicellular organisms are structures of single-celled organisms — and so on up, through botany, zoology, entomology, psychology, sociology, economics, etc. All these levels of structure may be called the **Grand Structure**, the product of ultimate inter-disciplinary scientific studies, a summary of the major discoveries and explanations of all the sciences, and the ultimate unification of science; as such it has to be important in any discussion of the noumenal world. Not only does this Grand Structure have levels of structure, but each level has novel emergent properties: properties that do not occur in the levels below. Atoms and molecules have chemical properties; single-celled organisms have life, which includes feeding on the environment and reproduction; multicellular organisms may have mobility, and nervous systems, and nervous systems may have brains which have minds and consciousness. All these emergent properties are parts of structures, and the Grand Structure is a structure of structures of structures, of many levels, and all these structures are relations relating relations. And the Grand Structure, in so far as it is noumenal, explains much of the empirical world. Hence much of the noumenal world consists of relations, and so, by induction, probably all of it does. And of course there is other structure in the noumenal world: hydrogen atoms form thermonuclear stars which explode and form atoms of higher atomic number which form suns with planets which have oceans and continents which have life.

This circular chain of consequences, from relations to mathematics, then to theoretical science, then to explanations, then to underlying causations, then to the noumenal/empirical distinction, then to the noumenal world, then to the noumenal consisting of relations: this circular chain is the main subject of this book. But since it is based upon replacing set theory with relations, as the foundation of mathematics, logic, and science, there has to be a fair amount of discussion of set theory, logic, and mathematics as well.

Because relations are primitive we cannot say what they are, we cannot define them, but they can be characterised; this is the subject matter of Chapter 1.

In Chapter 2 sets are defined and examined using relations as a foundation.

One of the many interesting consequences of making relations basic is the discovery of three different kinds of meaning of words and symbols, here called relational meaning, extensional meaning, and nominal meaning. **Relational meanings** are relations and properties of relations, and they include intensions of sets; **extensional meanings** are sets and their extensions; and **nominal meanings** include the words or symbols for all of these. Relational meaning allows axiom generosity (40) and is consistent; nominal meaning allows paradoxes and contradictions, and may therefore be inconsistent; and extensional meaning allows neither generosity nor inconsistency. The distinction between these three kinds of meaning is further discussed in Chapter 6.

We can apply the three kinds of meaning to the concept of **definition**, the most basic relationship between language and fact. **Relational definition** is definition of words that have relational meaning; it has the advantage of necessity. **Extensional definition** is definition of words that have extensional meaning; it has the advantage of universality, and it relates classifications, as in taxonomy. **Nominal definition** is synonymy, which may define nothing.

Given these distinctions of meaning it is possible to define three kinds of logic: relational logic, extensional logic, and nominal logic; these are discussed in Chapter 3. Three kinds of set theory — relational, extensional, and nominal — are discussed in Chapter 4. And relational mathematics, extensional mathematics, and nominal mathematics are discussed in Chapter 5.

Another interesting feature of the relational approach is a property that some relations have: the property of **extrinsic necessity**. We are familiar with it in logical necessity and mathematical necessity and we assume its presence in explanations. We believe in causal necessity because we need explanations. Explanation is further dealt with in Chapter 7.

The distinction between the theoretical and the empirical is in fact quite sharp, although most physicists do not acknowledge it. I once was told by a physicist that at the time that John Dalton (1766-1844) proposed his atomic theory, atoms and molecules were completely theoretical; but as time went on they became increasingly empirical until, with the scanning tunnelling microscope they became completely empirical. He was incorrect, of course: he was confusing evidence for the theoretical, which is empirical, with the theoretical itself, which is not empirical. What he thought of as 'becoming increasingly empirical' is the evidence for atoms, not the atoms themselves. Consider any microscope — optical, electronic, or tunnelling: what it shows is an enlargement of an object that is too small to see, but it does not enlarge the object — how could it? — it only enlarges an image of the object. The object and the image are two, not one. The object is invisible, the image is visible; the object is theoretical, the image is empirical. In everyday living this distinction is not important, but in science and in philosophy it *is* important. That is why, in this book, we distinguish two realms: the empirical world, consisting of everything that exists and is perceptible, and the noumenal world, consisting of everything that exists and is underlying, or imperceptible. This explains why there are two kinds of science: empirical and theoretical: empirical science describes the empirical world and theoretical science describes, or tries to describe, the noumenal world. So all causations are noumenal, and distinct from empirical correlations both in being necessitous and in being the causes of those correlations. And we postulate the noumenal world because we cannot manage without explanations — we want to know *why*. That the noumenal world consists of relations, and only relations is further discussed in Chapters 7 and 8, and the relations between the empirical world and the noumenal world are discussed in Chapters 11 to 13.

Chapter 1. Relations and their Properties.

We first need to resolve four difficulties that many people have with relations. They are: difficulties in ordinary language with relations; the perceptibility of relations; their often ephemeral existence; and the endless multiplication of some relations.

The difficulty that most people (to say nothing of philosophers) have with relations is shown by errors in ordinary language about relations. Anyone might say that their nearest relations (or relatives) are their parents, spouse and children; other relations being siblings, aunts and uncles, cousins, grandparents, in-laws, and step relations. This usage is quite incorrect logically, but it follows the common sense attitude of avoiding relations in language wherever possible. A woman's father is not a relation of hers; rather, she and he are both terms of the asymmetric relation called fatherhood in one ordering of the terms, and called daughterhood in the other. Common sense has trouble with this because fatherhood or daughterhood cannot be perceived and so seem to be somehow unreal, whereas a father and his daughter can be perceived and are concretely real. Many other relations in everyday experience are linguistically hidden, disguised in the form of transitive verbs: "Alice gave Eve a message for Bob" essentially describes a tetradic relation, *gave*, having the four terms Alice, Eve, message, and Bob. Note that the property of transitivity, as in 'transitive verb', is an abstract property of relations. Other examples of relations, here italicised, are "The judge *sentenced* Eve to three weeks", "You *are driving* too fast", "You *think about* it", "Two *plus* three *are* five", and "Bob *begat* Alice". Yet other linguistic ways of dealing with relations are metaphors and analogies, both of which convey meaning by similarity of their structure to their referents — and both similarity and structure are relations.

The second of the difficulties with relations is the question of whether or not they can be perceived. In fact they can be perceived because many of them exist empirically, but this contradicts the common belief that only the concrete can be perceived — the **concrete** being anything perceived or imagined other than relations. It follows that the **abstract** is anything that is not concrete — which is to say relations and their properties. These two definitions are stipulative in this book.

My favourite example of an abstract relation is a mug of coffee. You can see quite clearly that the coffee is *in* the mug; and this *in* is a relation, having the coffee and the mug as its terms. But although you can perceive the relation *in*, you cannot say what it looks like, or sounds like, or tastes or smells like — in which case it seems to be imperceptible: if it has no perceptible properties how can you perceive it? However if the relation *in* between the coffee and the mug were not perceptible, how could you know that your mug had coffee in it? And if the relation *in* were not real, how could you drink your coffee? So: the empirical coffee is *in* the empirical mug: the *in* has no concrete properties, so is an abstract entity, and has to be empirically real if you are able to see it, and able to drink the coffee. Empirical relations most often have concrete terms, either in the empirical world or, introspectively, in the imagination. And it is a simple fact that we perceive many empirical relations. Not only can we perceive abstract entities — relations — we can also perceive some of their abstract qualities. For example, I can perceive the *in* of the coffee to be asymmetric because although the coffee is *in* the mug, the mug is not *in* the coffee. And, if I put a spoon *in* the coffee, I see the *in* to be transitive: because the spoon is *in* the coffee, and the coffee is *in* the mug, it follows that the spoon is *in* the mug. Note that this asymmetry and transitivity are abstract properties, because they are not in any way concrete, and also that they are empirical. Note also that a concrete object may have abstract relational properties, as in the mug having a shape and in being topologically a torus because of its handle. Note thirdly that relations cannot exist without their terms, so if we did not perceive these empirical terms then we could not perceive the relations — and it is important that we perceive the relations because, among other things, some of them have survival value — as in the perception of danger: among the relations between a nearby loose tiger and you would be that of danger.

Yet a third reason for scepticism regarding relations is their seeming ephemeral nature. If you put your hat *on* your head this instance of the relation *on* comes into existence, and if you take off your hat this *on* goes out of existence — unlike your hat and your head, which endure. However not all relations are this ephemeral, but they do all have this characteristic of easily coming into, or going out of, existence. When a relation comes into existence it is said to **emerge**, and one that goes out of existence is said to **submerge**.

A fourth puzzle about relations is that some of them seem to multiply endlessly. Consider relations of self-similarity. We say that everything is self-similar simply because nothing is dissimilar to what it is: it may be dissimilar to what it was, or what it will be, but it cannot be dissimilar to what it is now. But if relations are real entities, even though abstract, then any instance of self-similarity is a real entity and so is self-similar; and this second self-similarity is a real entity and so itself self-similar, thereby producing a third self-similarity, and so on without end. This also applies to the relation of self-identity, if there is such a relation. This kind of multiplication is called **endless multiplication**. Another endless multiplication occurs with the claim that a relation *has* terms; possession is a relation, holding between possessor and possessed, so there seems to be a relation of possession between every relation and each of its terms, so each relation of possession has a relation of possession to each of its terms, leading to another endless multiplication of relations of possession. Another endless multiplication occurs if each relation possesses its properties.

1. Relations and their Properties

Another kind of unwanted multiplication is **extravagant multiplication**, which is multiplication that is less than endless multiplication. Any relation of similarity, S_1 (not to be confused with self-similarity) is itself similar, S_3, to any other relation of similarity, S_2, thereby promoting a huge number of relations of similarity. Equally, every proton is dissimilar to every neutron — and there are a vast number of those, and there are many, many other dissimilar things besides protons and neutrons in existence; and every dissimilarity is similar to every other dissimilarity; and every electron is similar to every other electron — and there are a vast number of electrons, and many, many other similar things.

So if relations produce this huge ontological clutter, can they be real? Many have thought not, including Spinoza, Leibniz, and Kant. But the answer is yes: many relations are real. Relations such as greater and less, equality, and spatial and temporal relations are real, but any that multiply needlessly are not. This is true empirically, since no one has ever perceived a huge multiplication of empirical relations. And it is true noumenally, by Occam's Razor: do not invent more theoretical relations than are needed to explain the empirical facts. There is also a principle behind this: it is that relations exist if they must, if they are necessitated; or, equivalently, they exist if something that does exist cannot exist without them. For example, a relation cannot exist if its terms do not exist. So if nothing needs the existence of a relation then that relation does not exist; and since nothing needs a huge ontological clutter, that clutter does not exist. Another example of necessity is the success of theoretical science, which necessitates the existence of the noumenal world, *via* the facts that explanation is causal and causation is imperceptible. Another criterion for the existence of noumenal relations is their explanatory value.

So we deny infinite ontological clutter: either by denying outright any relations that produce it, or else by putting restrictions on their existence. Thus we can claim that most relations with only one term, such as self-similarity and self-identity — monadic relations, as they are called — simply do not exist in empirical reality or noumenal reality. They are here called **exclusively nominal relations**, or, equally, **purely nominal relations**: they can be named and described, but the names and descriptions have no empirical or noumenal reference, no denotation.

Relations that multiply unnecessarily but must exist because of their explanatory value, such as relations of similarity and dissimilarity, are dealt with in Section 11 (32), where we deal with other relations that threaten to multiply endlessly, such as set-membership and the relation-every.

There is much that is unknown about relations. If you could discover these unknowns then you would be able to explain everything. There is also much that is known about relations; we go into this next.

We begin with two basic, or undefined, or primitive, concepts: **relations**, and their **properties**, and we take as understood two specific but familiar primitive concepts: *intrinsic* and *extrinsic*. Out of this come sixteen numbered Sections, each dealing with a particular feature of relations. Please note that symbols used herein are defined on page 45.

1.   This Section deals with intrinsic properties of relations and Section 2, below, deals with their extrinsic properties.

The **intrinsic properties** of a relation are what distinguish one **kind** of relation from another.

Intrinsic properties are here generally simply called **properties**, although they will be labelled intrinsic if it is needed to emphasise that they are this. Like their relations, they are primitive, they are not defined.

There are three properties that all relations have: **adicity**, **simplicity**, and **possibility**. The set (29) of these three is called the **minim**. Most relations have more properties than the minim, but we will deal with the minim first.

1. Relations and their Properties

The *adicity* of a relation is a primitive property that we call the number of its terms; so relations may be described as dyadic, triadic, tetradic, polyadic, etc.

Adicity is an *instance* of a number, *n*, say; this number itself is the intension of the set of all instances of a number that are equiadic (91) with *n*.

Adicity of relations, the number of their terms, makes relations fundamental in arithmetic and so in mathematics. In other words, given relations as primitive and basic in mathematics, it follows that number is also primitive and basic. (This concept of number is not to be confused with the concept of extensional number, defined set-theoretically (5, 98) and briefly discussed in the Introduction. The least possible adicity of a relation is two: there are no zero-adic or monadic relations: such "relations" are not relations at all (see Section 2 below) so are merely nominal.

The *simplicity* of a relation means that it has no parts, it is one, it is indivisible. Because all relations are simple this means that concepts of relations cannot be analysed into any simpler concepts. Relations are thus quite basic, and quite distinct from the logistically defined extensional relations that were mentioned in the Introduction (5).

There are four things to say about the *possibility* of a relation, the third property in the minim. It is first of all **consistency,** or lack of contradiction: a relation that is inconsistent is not possible.

Second, possibility is **mathematical existence**: if X is possible in a mathematical system then it exists in that system, and if X is not possible in that system then it does not exist in it. (Mathematical existence is also sometimes called **logical existence**; for ease of exposition it will hereafter be referred to simply as mathematical existence.)

Third, possibility may be singular or plural: if it is singular it is a **necessity**, and if it is plural it is a **contingency**. The distinction between these is shown in axiomatic systems: the choice of axioms is contingent, the theorems that result do so necessarily. This means that there are two kinds of existence: mathematical and actual, defined respectively as plural and singular possibility. Actual existence is the existence that we know empirically, the existence of the empirical world — and also the existence of the noumenal world. It may seem surprising that actual existence is defined in this way, but justification will be provided later (134).

Fourth, possibility transmits possibility, and does so necessarily. For example, a relation cannot exist without its terms, so if the relation exists then its terms exist necessarily. Equally, properties cannot exist without a relation, so if the properties exist then the relation exists necessarily. Such transmitted existence may be either mathematical or actual — contingent or necessary — but the transmission itself is necessary. We are familiar with such transmission in logical implication — e.g. in mathematical proof — and in genuine mathematical functions (111), in which the argument of the function necessitates the value — e.g. $\sin(\pi/2)$ necessarily is 1. We will come across other such necessary transmissions of possibility in due course.

We discuss possibility further in Section 16 (42).

Needless to say, there are other properties of relations besides the minim. Properties — often novel (21) — emerge from an emergence-configuration (20) of the terms of their relation.

2.   The **extrinsic properties** of a relation are either lower or upper extrinsic properties; they are what distinguish one **instance** of a relation from another. The **lower extrinsic properties** of a relation are its **terms**. The upper extrinsic properties of a relation R are all the relations of which R is a term; these are called the **uppers** of R.

All relations, and only relations, have terms. The terms of a relation R are other relations, except in the empirical world and in the imagination, where they may also be empirical things and empirical qualities; but we are primarily

concerned with terms which are relations, because we have
good grounds for claiming that the noumenal world consists of
relations relating relations, for many levels — as in the Grand
Structure (8); so we will assume that the terms of a relation are
other relations unless otherwise stated. All the terms of any
relation R form a set called the **term set** of R. The upper
extrinsic properties of a relation R form a set called the **upper
set** of R.

The terms and uppers of a relation determine the
instance of that kind of relation. Generally the terms are more
significant than the uppers in this respect. It is important (132)
to be able to be precise in the distinction between *kind* and
instance: the **kind** of a relation is determined by its intrinsic
properties, and the **instance** by its extrinsic properties. For
example, the kind of relation called fatherhood is
distinguished from the kind called motherhood by the intrinsic
property male parent, as opposed to female parent; and a
particular instance of fatherhood is the father of Bob and
Alice, distinguished by its terms, Bob and Alice.
Mathematicians are much more interested in kinds than they
are in instances; and, more generally, it is recognised that
definition should be by kind since definition by instance — by
example — although often helpful, is inadequate. The
definition of number by means of adicity is a definition by
kind and the modern set-theoretic definition of number —
extensional number — as outlined in the Introduction, is by
totality of instances.

If a relation R is a term of relation S then the set (29)
consisting of S and its terms, other than R, is an upper of R;
these other terms of S are called the **coterms** of R in S.

A relation R may have more than one upper, or none.
For example, people may be terms of relations, so a particular
woman, Alice, say, may have the uppers of a husband, Bob,
and a child, Eve; or, less colloquially and more accurately,
Alice has the coterm Bob in an instance of the relation of
wedlock, and the coterm Eve in an instance of the relation of

parenthood. These particular instances of wedlock and parenthood are dyadic, but other relations and adicities are possible: Alice and Bob might also have children Jack and Jill, so that Eve would have Jack and Jill as coterms in siblinghood.

The terms of a relation have a configuration. This is defined later (39) but for now may be understood as one of two kinds: **emergence-configurations** and **submergence-configurations**. If the terms of an instance of a relation R exist and have an emergence-configuration then R has to exist, and if the terms exist and have a submergence-configuration then R cannot exist; thus an emergence-configuration necessitates the existence of a relation. Submergence is absence of emergence, so is only a convenient word that has no denotation: absence is non-existence, so has nominal meaning only.

There are three necessities with relations: one between a relation and its terms, another between an emergence-configuration and a set of properties, and a third between a set of properties and a relation. These necessities are called respectively **demergence**, **emergence**, and **inherence**, and together they form a loop, called a **minor loop**. A relation, R, demerges the set, T, of its terms (if R exists then its terms have to exist), the terms form an emergence-configuration, C (one out of many possible configurations; what determines C is necessitated from outside the loop) and C necessarily emerges the set, P, of intrinsic properties of R, and these properties (which cannot exist without R) inhere the existence of R. If we symbolise demergence by ⇓, emergence by ⇑, and other necessities by ⇒, then the loop is symbolised by R⇓T⇒C⇑P⇒R (the first R being identical with the last R), with the understanding that C is partially necessitated from outside the loop. Also, R has one or more uppers which are parts of configurations at a higher level, and which demerge R, so the loop is necessitated from above as well as from below. We go into more detail on this in Section 3. Note that, when writing loosely for simplicity, we may use the expressions 'the emergence-configuration emerges R' or 'R emerges from an emergence-configuration' this is in fact an over-simplification: rather, the emergence-configuration emerges *P* which inheres R.

Emergence is **bottom up**, or **upward** necessitation, and demergence is **top down**, or **downward**, necessitation. These directions are from lower to higher levels (24) or *vice versa*.

An important characteristic of emergence is that an emergence-configuration may emerge one or more novel properties. This is very common in mathematics. A property is **novel** in the sense that it does not emerge at any lower level; the lowest level in which it does emerge is known as its **emergent level**. An example is an old-fashioned weight-driven pendulum clock, such as was invented by Christiaan Huygens (1629-1695), which has the emergent property of telling time, while none of its parts can tell time. In fact any machine has an emergent function, for which it was designed, and which is not present in any of its parts; any **design** is a process of making wholes (26) with specific emergents: this includes architectural design, design of towns, furniture, machines, food, clothing, etc. Another example of novel emergence is knots: as any sailor or scout knows, there are a wide variety of knots, emergent out of loops and threadings of cord or rope, each of which has properties not possessed by the cord, rope, loops or threadings: a reef knot is easily undone, a bowline will not unravel, and so on. A knitted sweater is a complicated knot having the designed emergent property of keeping you warm. Yet another example of novelty is a melody, emergent out of a particular temporal configuration of notes but not out of other temporal configurations; and these notes are not themselves melodious. And a computer programme is a novel emergent out of machine language; in fact, there are many levels of language in computers: machine language, assembly language, and various levels of programming language. The best of all examples of emergence is the Grand Structure (8).

We can define the novelty of novel emergent properties: it is dissimilarity between the emergent property and all lower level properties. And we can offer a partial explain why novelties must emerge: a dissimilarity demarcates, it is a boundary; so if, for some reason, there have to be boundaries between levels then novelties are necessary.

The most puzzling feature of relations is the fact that we have no idea what it is about configurations that makes them emerge novel properties, nor what those novel properties will be. Given a set of properties and its emergence-configuration, we cannot predict what novelties will emerge.

3.　　　　A special kind of relation is a **necessity relation**. It has one term, or a set of terms, called the **antecedent**, and another one term, or a set of terms, called the **consequent**; and the antecedent cannot exist without the consequent, so that he mathematical existence of the antecedent necessitates the existence of the consequent. This necessitation is an intrinsic property of the necessity relation. We see this with the minor loop $R \Downarrow T \Rightarrow C \Uparrow P \Rightarrow R$, where each of the arrows $\Rightarrow$, $\Uparrow$, and $\Downarrow$ is a necessity relation. This is illustrated in more detail in Fig. 1.1.

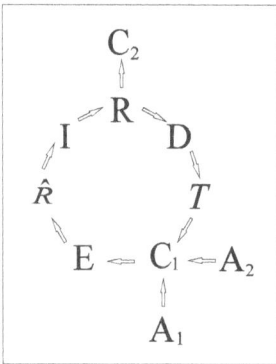

Fig. 1.1

Here the arrows show only the asymmetry of the various relations. The relation R cannot exist without its term set, T, so R demerges (D) the term set T. T cannot exist without a configuration, C_1, so C_1 is partially necessitated by T; C_1 is also partially necessitated from outside the loop, by antecedent A_1 and/or A_2: A_1 is a lower level antecedent and A_2 is a cause. C_1 emerges (E) the property set $\hat{R}$, which cannot exist without the relation R, so $\hat{R}$ inheres (I), the relation R. (We also say that R exheres r̂; exherence is the inverse of inherence.) R then becomes a member of C_2, a higher level configuration, just as does each member of *T*, form a lower level configuration. Thus all non-necessity relations, such as R and members of *T*, both emerge from below and demerge from above; the loop relations D, E, and I do neither: their necessity is an intrinsic property of each of them.

Thus demergence, D, inherence, I, and emergence, E, are all necessity relations. Other necessity relations are causations, genuine mathematical functions (110), and implications. We see the necessities in the latter three in that a cause cannot exist without the existence of its effect, an argument of a genuine function cannot exist without the existence of its value, and the antecedent of a true implication cannot be true without the truth of the consequent. Necessity is a special case of possibility (42) and possibility is an intrinsic property of a relation — it's in the minim — so all these necessities must be intrinsic properties of necessity relations;

Although a necessity relation may have many terms, many properties, and many uppers, each of these may be treated as a set (3, 35), and so one term. So necessity relations seem to be all dyadic. However, as our discussion of sets (3, 35) shows, this is an error. A causation, for example, may be polyadic because of having many antecedents; each antecedent is a partial cause of the consequent. Another error in popular usage is to say that relations emerge from their terms; as we have seen, relations are inhered by their properties, which emerge from emergence-configurations of their terms; however this simplified usage is harmless in everyday speech.

Necessity relations are also all asymmetric: antecedent precedes consequent. And a necessity relation transmits possibility: the possibility of the antecedent necessites the possibility of the consequent — and possibility is mathematical existence, so in a minor loop, such as that of Fig. 1.1, existence necessitates existence.

4.　　　We define the **level** of a relation R as: If R has level-*n*, where *n* is a number, then the terms of R are at level-(*n-1*), the next lower level, and the uppers of R — the relations of which R is a term — are at level-(*n+1*), the next higher level. The lowest level of a structure (8, 25) is level-*1*, also called the **axiom level**, and it consists entirely of relations called **separators** (39); at this level separators are also called **axiom relations** (30). The highest level is called the **top level**; it must consist of a single relation, since each level has fewer relations than in the level below because every relation (except axiom separators) has at least two terms in that lower level. Nothing can emerge from the top level because emergence is from an emergence-configuration and a single relation cannot constitute a configuration.

5.　　　Whereas empirical things and empirical qualities are concrete, relations are abstract, meaning that relations are not concrete and they have no concrete properties — although empirical relations usually have concrete terms. The fact of being abstract is not a property of relations because it is merely the absence of concreteness, and absence of a property is not a property, except nominally.

　　　The distinction between the concrete and the abstract is important in the present context. The **concrete** is any empirical quality, any quality known through the senses — since this is what 'empirical' means. Examples are tactile sensations: various degrees of hard and soft, rough and smooth, hot and cold, wet and dry, large and small, heavy and light, and degrees of solidity, plasticity, slipperiness, ductility, fragility, etc.; and a huge range of colours and sounds; and tastes and smells. The concrete is also any structure (8, 25) of concrete qualities, which includes any empirical object.

　　　The **abstract** is anything that has no concrete properties: relations and their properties are abstract, and in the present work the word abstract is confined to these.

Also in the present work, **imagination** is the part of the mind dealing with the concrete while **thought** deals with the abstract; that is, imagination works with images of the concrete, and thought with abstract ideas, which are relations existing in the mind. In ordinary language a word such as 'dog' is considered to be abstract, in that it refers to all dogs or to any dog; but here it is considered simply a name for the extension (3) of the set of all dogs. Note that both perception and imagination may contain relations: relations having as their terms concrete qualities and/or concrete objects. The definition of the abstract given earlier, in the Introduction, was that it is anything that cannot be perceived or imagined in isolation; this is equivalent to saying that it has no concrete properties, since concrete properties *can* be imagined in isolation; and relations cannot be imagined in isolation, because a relation cannot exist without its terms, and intrinsic properties of relations cannot exist without their relations.

6.   Levels of relations develop when relations relate relations, as we have seen with the Grand Structure and in Section 4. Out of the concept of level we can define a structure, a whole, and a possible world.

A **structure** is a set of levels of relations, starting in level-*1*, the axiom level; each level has fewer relations than the level below, because each relation must have at least two terms in the level below[2], so the structure ends in a top level consisting of a single relation, called the **top relation** of the structure. The top relation may itself be a term of a relation in the next higher level, so that a structure may be a part of a larger structure. Each level of a structure is emergent from the previous level; such successive emergences out of the axiom level is called **cascading emergence**. We are familiar with it

---

[2] Some relations may share terms: a change, for example, is a dissimilarity and a duration with common terms. But these are rare so do not invalidate the argument.

in the Grand Structure and in the cornucopia of mathematical theorems that emerge from a good axiom set.

A structure which has a novel emergent property in its top relation is called a whole, as opposed to a mere structure. A **whole** is defined by an emergent top relation, T, which possesses at least one novel property — a property that does not appear on any level lower than that of T. The terms in lower levels of T are together called the **parts** of the whole if they are top relations of wholes, while they are otherwise called **subordinate terms** (90) of T. The novel property or set of properties that define a novel kind of relation, and hence a whole, is what makes the whole greater than the sum of its parts.

A **possible world** is a special case of a whole in that its top relation is not a term of any other relation: it does not have any uppers, it has no coterms. It is important to understand why this is so. It is because a single relation cannot constitute an emergence-configuration, so nothing can emerge from it. So a possible world has to be complete. Also, a possible world is a candidate for being a true description of the noumenal world, and the noumenal world is defined as *everything* imperceptible that exists; the noumenal world is complete by definition, so a true description of it must also be necessarily complete.

So a whole is not necessarily a possible world, since it may be related to other wholes by a relation that has their top relations as its terms. And a structure is not necessarily a whole, since a structure does not necessarily have novel properties in its top relation. We examine possible worlds in Chapter 8.

7.   There are three important categories of relations: **empirical**, **ideal**, and **noumenal**. Empirical relations are relations that we each perceive in the empirical world around us, or introspectively in the imagination. Ideal relations are relations in thought. Noumenal relations are relations that exist

but are neither empirical nor ideal, such as non-empirical relations in the Grand Structure. As we saw earlier, the totality of empirical things, qualities, and relations, other than those in the imagination, is called the **empirical world** and the totality of noumenal relations is called the **noumenal world**.

Empirical relations are relations that are known through the senses, as the word empirical means; they have as their terms empirical things, empirical qualities, or other empirical relations. Otherwise all that needs to be said here of empirical relations is to provide some examples. Empirical spatial relations include *near* and *far*, *high* and *low*, as well as *direction, between, beyond, behind*, and *beneath*; out of these emerge *shapes, sizes*, and *greater* and *less* Empirical temporal relations include *earlier* and *later, younger* and *older, duration, simultaneity*, and *between*. Spatial and temporal relations, suitably related, together yield *velocity* and *acceleration*. Empirical *correlations* may be evidence of causations, but are not actual causations because there are no necessities between the correlates.

Ideal relations are relations in the mind of a thinker — the word ideal is here used in the philosophic sense of consisting of ideas, not in the Platonic sense of perfection. Ideal relations include instances of any of the empirical kind of relations, as well as any of the relations of logic and mathematics, including those of theoretical science. But in this last realm there is a problem: the problem of ignorance. As science progresses our ignorance diminishes, but we are ignorant of the extent, and content, of our ignorance. This in turn leads to the concept of **possible noumenal worlds**. If we do not know the full details of the actual noumenal world then we have to be able to consider various possible noumenal worlds, and compare them — with each other and with the empirical world. But the actual noumenal world is the totality of noumenal relations, while a comparison of two or more possible noumenal worlds is a relation outside that totality — as well as the fact that two or more such totalities cannot co-exist, by definition. So we distinguish two further kinds of ideal relations: **worldly relations** are ideal relations *within* a possible noumenal world — that is, an ideal possible world — and **exotic relations** are ideal relations *between* possible noumenal worlds. Exotic relations cannot be noumenal, by relating two or more noumenal worlds, because the noumenal world is the totality of noumenal relations and so is the one and only noumenal world; but they are necessary while we try to cope with our ignorance of the noumenal. From now on reference to possible worlds is reference to ideal possible noumenal worlds, while reference to the noumenal world is usually reference to that possible world which describes, somewhat, the real noumenal world. Also, any complete mathematical system is a possible world, emerging out of an axiom set.

David Hume (1711-1776), who said that there are no necessary connections between correlated events, also said that 'tis vain to speculate. He was writing as an empiricist, for whom all knowledge is empirical, hence non-empirical things cannot be known so 'tis vain to speculate concerning them. If he was correct in this then nothing noumenal can be known, in which case nothing theoretical can be known. What Hume missed was that if 'tis vain to speculate then 'tis vain to try to explain. Explanation is in fact the key to all noumenal claims. The criterion for qualification as noumenal is **explanatory value**: the better a theory is at explaining the empirical, the more probable is the noumenal existence of its content. **Noumenal existence** is actual existence in the noumenal world: singular possibility, as opposed to mathematical possibility, which is plural. The criteria of good explanation are discussed in Chapter 7.

What might be called a fourth category of relations is purely **nominal relations**: relations which exist in language, as names or definitions, but the language has no reference, no denotation. Since these relations do not exist other than nominally they are of little interest, but we do need to be able to speak of them, if only to dismiss them in order to correct error. Extensional relations (5) are in this category.

8.   Relations in the mind form **natural sets**: every relation has a set of intrinsic properties, called a property set (45); and a set of lower extrinsic properties called a term set (19), as well as, usually, a set of upper extrinsic properties, called an upper set (19). These three kinds of sets are natural sets; they are also **intensional sets**, which are sets which have intensions, the **intension** of a set being that property or set of properties that all and only the members of that set have. For example, the term set of a relation R is every term of R; so if something is a term of R then it is a member of the term set, and conversely if it is a member of the set then it is a term of R. However an intension by itself does not define a set: also needed is the self-explanatory **relation-every** (46), symbolised by $\mathcal{E}$; thus if p is an intension then $\{\mathcal{E}p\}$ is the intensional set defined by p. $\mathcal{E}$ is a relation between intension and set.

Intensional sets are also called **necessary sets** because their membership is necessitated by their intensions and the relation-every.

An intensional set may be defined by enumerating all of its members, if this is possible, or by its intension and the relation-every.

We later will also consider **contingent sets** which are sets that do not have intensions and so can only be specified by enumerating their members. They are called contingent sets because they are not necessary sets. And we will consider **extensional sets**, which are sets which are either necessary or contingent.

By analogy with all these ideal sets we may talk of sets of empirical things and sets of empirical qualities.

We may also define **nominal sets**, which are names or descriptions of sets; some of these have reference to intensional sets, to contingent sets, or to extensional sets; and others — purely nominal sets — have no reference at all, such as the set of all even prime numbers greater than two, but such a fictitious set can still be described or named: it is called a **null set**, symbolised by ϕ.

A more detailed account of sets is given later, in Section 17 and in Chapter 2.

9.   The concept of **separator** (more precisely defined later (39)) arises from the fact of relations relating relations, which produces levels. Because the variety (132) of relations must decrease with level, there must be a lowest level, level-*1*, already named the axiom level, in which variety is a minimum. At this level are relations which both have similar relations as terms and are themselves such terms, all on the same level. These relations are called separators because separators separate separators and are themselves separated by separators. Consider a chain of hyphens:

$$\ldots -\ -\ -\ -\ -\ -\ -\ldots$$
$$\quad\ \ \text{A}\ \ \ \text{B}\ \ \ \text{C}\ \ \ \text{D}\ \ \ \text{E}$$

where the subscripts A, B, C, etc. are merely names of the hyphens and not part of the chain. This shows that hyphen C separates hyphens B and D, so C is a relation, a separator. But C is itself a term of B because B separates A and C and so B is a relation; and equally C is a term of D because D separates C and E. Thus B and D are both uppers of C and also terms of C. This is true of all axiom-level separators, making each of them relation, terms, and uppers, all on the same level. They may be thought of as quanta: a linear separator is a quantum of length, a temporal separator is a quantum of duration, and so on. They may also be thought of as atomic lengths, atomic durations, etc. This is not only because separators, like all relations, are indivisible, but also because they are compoundable relations (38); being compoundable means that, among other things, they may also exist on levels higher than the axiom level.

Axiom separators are relations that do not emerge: there are no emergence-configurations for them to emerge from because there is no lower level; but they are demergent from above.

It is very probable that separators are Planck units (see Appendix D).

10.   It is characteristic of axiom separators that they form **chains** and that these have to form **closed loops**, called **separator loops**; if the chains did not form loops then they would have end separators and these would each be missing a term and so could not exist, in which case the next separators would be ends and so could not exist, and so on through the whole chain.

The links in these loops do not have to all be similar: if there are other possible links — separators — which are compatible with the main links in the chain, then these other possible links may appear in the chain. When this happens there is a dissimilarity in the chain and this dissimilarity is a boundary within the chain. (This dissimilarity is a separator but limited because it does not have uppers (33) so does not form loops.)

There are other loops in possible worlds; they are of three kinds, **minor loops**, **middle loops**, and **major loops**. A minor loop occurs with minimal emergence and demergence: as we saw above (20, 22), $R \Downarrow T \Rightarrow C \Uparrow P \Rightarrow R$. R and C are also parts of higher and lower level loops, so minor loops form chains of loops and thus form links in a larger loop. When such a larger loop is a whole it is called a middle loop; the top relation, T, of the whole demerges minor loops all the way down to, and including, the axiom level; and the axiom level loop cascadingly emerges all the minor loops up to, and including, T. A major loop is a middle loop that includes the top relation of a possible world (27).

The two key features of these loops are that what is being looped is possibility, which is mathematical existence; and the links are necessities. So the loops are necessary chains of mathematical existence, or **loop possibility**, which is to say chains of singular possibilities of possibility.

The significance of loops is discussed in Chapter 8.

11.   We next deal with the unnecessary multiplication of some relations. We earlier (14) distinguished two kinds: endless multiplication, and extravagant multiplication. A relation that multiplies endlessly, such as self-similarity, cannot exist — either empirically or noumenally — although it may exist ideally. A relation that multiplies extravagantly, such as a similarity or dissimilarity, may exist empirically if it can be perceived, or noumenally if it has explanatory value, but only if its multiplication can be controlled; and even then its multiplication may be extravagant when ideal.

The ideal case is special because a mind may make comparisons (34) that threaten endless or extravagant multiplication, but not actually produce it. Anyone can compare two similarities and find them to be similar, and then compare this third similarity to the first two, to get two more, and go on multiplying similarities in this way as long as they like. This is not extravagant multiplication because no one chooses to do this for long. It is like thinking about thinking, then thinking about thinking about thinking, then thinking about thinking about thinking about thinking — no one goes far in such a progression. Furthermore, a mind can compare something with itself and discover it to be self-similar, and then discover this self-similarity to be self-similar, and multiply these as far as it likes; but this is not endless multiplication because the comparisons have to be made in order to produce the next self-similarity, and no one goes very far with this. Thus some self-similarities, and hence monadic relations, may emerge in minds; but they have no explanatory value, have ideal existence only, and may apply to everything ideal. And not only that, they are illusory because their one term does not form a configuration so they cannot emerge, yet they do emerge, from a comparison. They may be thought of as the mind spinning its wheels, like a car spinning its wheels when going uphill on ice. We acknowledge the possibility of such ideal relations, but from now on ignore them.

11.1   Self-similarity is the paradigm case of endless multiplication. As we saw earlier (14), we want to say that everything is self-similar because nothing is dissimilar to what it is. So self-similarity is a monadic relation, S_1, which like everything else, is self-similar; this second self-similarity, S_2, is also self-similar, thereby producing S_3, and so on, without end. And everything that exists is self-similar, hence starts such an unending chain. On the other hand, we do believe that nothing is dissimilar to what it is, hence is self-similar. We will resolve this paradox shortly (32), but for now simply claim that there are no relations of self-similarity because they multiply endlessly and explain nothing.

11.2   Similarity (not to be confused with self-similarity) and dissimilarity are paradigm cases of extravagant multiplication: they are relations that multiply extravagantly, but they do have explanatory value so we cannot simply deny them, as we do with self-similarity, so we control their multiplication by restricting their existence. For example, we need dissimilarities to distinguish one kind of relation from another, while a **boundary** may be defined as a series of contiguous dissimilarities, and a **change** may be defined as a necessary dissimilarity in parallel with a duration; and boundaries and changes clearly exist in the empirical world, so **dissimilarities** must exist. But since a dissimilarity would be a relation that would multiply extravagantly, we restrict its existence both by not allowing it to have uppers (it is never a term of another relation, and so has no coterms) and by allowing it to have only locally separated terms (its terms are spatially and temporally adjacent).

We could next simply define a similarity as the absence of a dissimilarity, and the absence of a relation is not a relation, so similarities would then not exist. On the other hand, if similarities have explanatory value and so may be declared to exist, then we can impose the same restrictions on their existence: they have no uppers and only locally separated terms. And they do have explanatory value: forces of attraction and repulsion in electrostatics and magnetism are based on like and unlike: like repels like and attracts unlike. And, of course, 'like' means similar and 'unlike' means dissimilar. If similarity was merely the absence of dissimilarity then we could explain attraction but not repulsion. So we must allow the existence of **similarities** at low levels. Further confirmation of this is the fact that all living things need to feed on their environment, and in order to do so they need to discriminate between edible and inedible. Explanation of this discrimination is by relating the questionable food to a standard: if it is similar to the standard then it is edible, and if it is dissimilar then it is inedible. Obvious examples of similarity also occur in mathematics, as with congruence, isomorphism, and the "unchanging changes" of group theory. So we limit similarities as we do dissimilarities: they are never a term of another relation, and so have no coterms, they have only locally separated terms.

It may be, of course, that a relation seems to multiply uselessly but we do not know what restricts it from doing so in reality. But this does not matter. It is enough to know that such multiplication is impossible and so must be restricted.

A possible source of confusion arises from the fact that similarities and dissimilarities also occur in higher level systems such as minds and computers, where comparisons produce them. Recognition, for example, results when a comparison of a present perception with a memory produces a similarity, which is a recognition. But these higher level similarities and dissimilarities are not limited in their existence as are low level similarities and dissimilarities. A mind can compare similarities by making comparisons, but it is only spinning its wheels.

11.3   More endless multiplication arises with *sets*, with the *relation-every*, and with *set-membership*. It is very tempting to say that a set is a relation, all the terms of which are the members of the set. But if a set is a relation, S_1, then it has a term set, S_2, and a property set, S_3; S_2 is then a relation that has a term set, S_4, and a property set, S_5, and S_3 is then a relation that has a term set, S_6, and a property set, S_7; and so on, endlessly. The same is true of the relation-every, which has a property set and a term set which can only be defined with the relation-every; and also the relation of set-membership has a property set and a term set which both have the relation of set-membership to their terms. So if sets are relations then they multiply endlessly, and the relation-every and the relation of set-membership do so whatever sets are. But if a set is not a relation, then what is it? The answer is that it is primitive, undefinable, like the concept of relation; thus it is ideal only, since endless multiplication disallows its empirical or noumenal existence; in ideal cases, endless multiplication is not a problem, as we have just seen. Equally the endless multiplication of the relation-every and of set-membership are ideal only. So we have to say that sets and the relations of set-membership and the relation-every are *entia rationes*, things of the mind, ideal only: they do not exist in either the empirical world or the noumenal world. And in fact this is not a problem. Whenever we find we want to use the word *set*, we can substitute the word *number*: a set of quaggas, for example, is simply a particular number of quaggas. This is especially clear with term sets: the number of members, *n*, of the term set of a relation is the adicity, *n*, of that relation. So when inventing theories we can define a **set** as a number of entities, and, given relations, this is no problem since every relation has an adicity and so numbers exist. In fact it is obvious that both sets and relations are characterised by numbers, so that it is easy to define a set in terms of relations or, alternatively, relations in terms of sets; and then the choice between these two approaches is determined by preference for one of the definitions of a number that they yield: adicity, or the set of all the sets that are in one-to-one correspondence with a given set

— although this choice is also influenced by the fact that relations have empirical and noumenal reference while sets do not.

It might be asked why it is that if sets and their associated relations are ideal only, why do we have these ideas? The answer is that we are born to classify (189). Sets are classes, and classes are the result of classification. Humans, and other animals, are born to classify, for a very good reason: classification has survival value: we classify the edible and the inedible, the dangerous and the safe, the valuable and the valueless — and survival is our strongest imperative. This connects with the discussion of ideal similarity, above, since most classification is on the basis of similarity. So we can say that sets, set-membership, the relation-every and the relation-any have no more than ideal existence; as such they are little more than the mind spinning its wheels, the 'little more' being our need to classify. We can allow this need, while recognising its inapplicability in science and mathematics.

Notice, however, that among all the concepts of set theory, the concept of intension usually does have real reference, since an intension is a relation plus an upper of that relation. For example, the intension of the set $\{\mathcal{E}\aleph R\}$ — the class of everything similar to R — has the intension 'similar to R', and R is a relation having similarity as an upper; the set $\{\mathcal{E}\aleph R\}$ has ideal existence only, but R and the relation of similarity may have real reference. R may also have a boundary as an upper, to yield a set such as 'The number of things in this box', where R is the relation *in* and the box is the boundary; the *in*, the box; and the number of things in the box, all have empirical reality, but the set and the relations of set-membership and relation-every do not.

So whenever set language is used in what follows it must be understood that (i) it is used because we are born to classify, and (ii) it is used in the sense of number rather than set.

1. Relations and their Properties

Finally I should like to emphasise that although this book demotes set theory in favour of relations, it is not a condemnation of the past century or so of work on the foundations of mathematics; it is simply an improvement of it.

12.　　　Relations have an **absolute value**, called their **hekergy**[3], defined as: if the number of possible emergence-configurations of a set of terms from which a relation R emerges is e, and the number of possible submergence-configurations of these terms is s, and $t = (s+e)$, then the hekergy of R is $\ln.(t/e)$. The numbers t and e are later shown to be finite. The ratio t/e is a degree of possibility and e/t is a probability (44); however although e/t is a mathematical probability, it has nothing to do with equiprobability or statistical frequency; whether a relation emerges or not depends only on the configuration of its terms.

We can perceive hekergies, but only subjectively — giving us subjective values rather than absolute ones, for reasons to be explained in Chapter 11. These subjective values include the traditional truth, beauty, and goodness.

Hekergy may be used to define the **quality of a possible world**. If a possible world contains s separators in its axiom level and has a total amount of hekergy, h, cascadingly emergent out of this axiom level, then the quality, q, of this world is defined as $q = h/s$. Note that axiom separators do not have any hekergy because they do not have configurations of their terms.

Given the definition of the quality of a possible world it follows that the ideal set of all possible worlds is a partially ordered set (a poset) because each possible world has a number, its quality, from which it follows that there is at least

---

[3] This word was invented for me by Prof. David Gallop at the University of Toronto, when I was an undergraduate there. It is based upon the concept of free-energy, from physics, that I was trying to generalise. I would now describe it as meaning a generalisation of negated entropy.

one highest quality world, one **best of all possible worlds**. We shall see that in fact there can be no more than one such best, and that it is a whole which has a top level higher than any other possible world, so that its top relation has a novel property unique among all possible worlds. This is discussed in Chapter 8.

A feature of hekergy of particular interest to physicists is that hekergies are a maximum when $e = 1$; so in the best of all possible worlds it is to be expected that stationary principles apply — principles such as those of paths of least action, shortest time or distance, or least resistance — since such paths are unique, hence have $e=1$ and so have maximum hekergy, which must exist in the best of all possible worlds. This, and other connections to physics, are discussed in Chapter 10.

13.   A **compoundable relation** is any kind of relation that has one or more properties which are also possessed by a **compounded relation**, R, which is a relation whose terms are a number of instances of that kind. For example, we may define an **atomic length** — a separator — such that a chain of atomic lengths has a compounded length. The compounded length is a relation having the atomic lengths as terms; as a relation it is thus simple. **Atomic durations** may be compounded similarly. A process is a temporal sequence of causations, each of which necessitates the next, so that the first transitively necessitates the last. Compounded relations are chains of similar compoundable relations, the chains being bounded by dissimilarities. More generally, all transitive relations are compoundable relations, such that a compound of them is of the same kind.

Compounded relations may multiply excessively if we are not careful. For example, an indefinite length composed of atomic lengths may have innumerable indefinite compounded lengths picked out along its total length. We deny this by claiming that only **definite compounded relations** exist, these being defined as chains of similar compoundable relations that have a dissimilarity at each end; these dissimilarities are the boundary of the compounded relation. Thus a compounded length consists of every atomic length between two dissimilarities. (Indefinite, or arbitrary compounded lengths may be selected by a higher level system such as a mind or a computer; this **selection** is an operation (that is, a relation) like a comparison, and although it produces multiplications, they are neither endless nor extravagant.)

An important kind of compounding occurs with dissimilarities and similarities, leading to an emergent **degree of dissimilarity** and an emergent **degree of similarity**. Given any two relations, if they have a number, d, of dissimilar properties, and a number, s, of similar properties, and one has m more properties than the other, then the **degree of dissimilarity**, D, between them is $D = (d+m)/(s+d+m)$; and the **degree of similarity**, L, between them is $L = s/(s+d+m)$. So if $D = 1$ then $s = 0 = L$, and if $L = 1$ then $d = m = 0 = D$; each varies between 0 and 1; and $D+L = 1$. We may in the same way define further degrees of dissimilarity and similarity between structures and hence between wholes and between worlds.

14.   A particularly important compounded relation is a configuration, a concept that can now be defined more precisely. If an emergent relation R has a term set R then the **configuration**, C, of R is the compounded relation consisting of the set of all of those compoundable relations whose terms are members of R — all the compoundable uppers of the terms of R, uppers whose other terms are all terms of R. In other words, each term of R has coterms that are other terms of R, and all these coterms are related by separators; these separators are the configuration of the terms of R. So the configuration of the terms of a relation R is all of the separators between those terms. As we saw earlier (20), with some configurations of the terms of a relation, the relation emerges, and with others it

submerges; so as configurations change, a relation may emerge or submerge.

15.   Relations and their properties constitute a special kind of meaning in language, here called **relational meaning**, which is distinct from **extensional meaning**, based on sets or classes, and **nominal meaning** which is language itself. The most cogent distinctions between these is that only relational meaning may have axiom generosity, only nominal meaning may be nonsensical, and extensional meaning has neither of these characteristics. These three kinds of meaning, all of which are significant in explanation, are the subject matter of Chapters 2 to 6.

        Axiom generosity occurs in mathematics, when a large output of theorems emerge out of a small set of axioms. The classic example of this is Euclid's *Elements*: five books that emerge from five axioms. Another example is the concept of right triangle, from which emerge Pythagoras' theorem, trigonometry, co-ordinate geometry, the complex plane, vector algebra and vector calculus, etc. A third example is any possible world, which emerges from its axiom level. The key point here is emergence: only relational properties emerge, axiom generosity is emergence, so only relational meaning has axiom generosity.

        (In the empirical world qualities emerge and submerge, as well as relations. For example, at dawn red emerges in the sky, then submerges, followed by the emergence and submergence of yellow, and then the emergence of blue. But it will be claimed later (185) that the emergence and submergence of concrete empirical qualities is caused by underlying emergence and submergence of relations: colours, for example, emerge because of the emergence of noumenal retinal excitation by electromagnetic radiation in the visible range.)

        Extensional meaning is reference to sets, either necessary sets or contingent sets, and to other features of sets such as intensions, extensions, and set-membership. Thus extensional meaning is mostly ideal only, but has real reference when the number of members of a set is applicable to reality; and the intensions of necessary sets usually also have real reference.

Nominal meaning puts relational or extensional meaning into words or symbols; but it may be only meaning by verbal analogy to either relational or extensional meaning, or meaning by definition of non-existent — impossible — things, and so have no reference; so as such it occurs only in language. For example, we can define *triangle*, *equiangular*, and *right angle*, and then define an *equiangular triangle*, a *right angled triangle*, and an *equiangular right angled triangle*. The first two have relational meaning if we are talking about the Euclidean plane, but the third does not; but having just written 'an equiangular right angled triangle' gives it nominal meaning — by verbal analogy to the first two definitions, which do have relational meaning.

It is a logical fact that, in language, relational meaning entails extensional meaning which entails nominal meaning. That is, relational meaning is a sufficient condition for extensional meaning and extensional meaning is a sufficient condition for nominal meaning, but nominal meaning is only a necessary condition for extensional meaning and extensional meaning is only a necessary condition for relational meaning. Thus a relational meaning, as an intension, entails an extension and both may be named or described; but a name or description may have no reference, and an extension may have no intension. This sequential entailment develops out of the property of necessity; it will be referred to sufficiently often in the following that an abbreviation for it is needed. This will be **R⇒E⇒N**. The three kinds of meaning are further examined in Chapter 6. (Meaning is, of course, not only a property of language, but the whole point of language. And meaning is not limited to these three kinds: there are moral and aesthetic meanings, for example.)

Two specific examples of the three kinds of meaning are **conjunction** and **disjunction**. A conjunction is denoted by the word *and*, and symbolised by ∧, and a disjunction is denoted by the word *or*, and symbolised by ∨. A conjunction is a compoundable dyadic relation whose terms are called **conjuncts**, and a disjunction is a compoundable dyadic relation whose terms are called **disjuncts**. The relational meaning of a conjunction is a coupling (73) of intensions, its extensional meaning an intersection (74) of extensions, and its nominal meaning is a join of words which is true only if all of its conjuncts are true, a **join** being a relation that has only the minim, $\hat{M}$, as its property set — the simplest possible property set — consisting of the properties of possibility, simplicity, and an adicity. The relational meaning of a disjunction is a commonality (73) of intensions, its extensional meaning a union (73) of extensions, and its nominal meaning is that it is a join of words which is false only if all of its disjuncts are false; this latter class may include **disjoins** (42), each of which is the absence of a join.

16.   We next consider the concepts of **possibility** and **necessity**. There are in fact three usages of these two words, which need to be sorted out.

The first usage is **logical possibility**, which is the primitive property of possibility that is in the minim and is also mathematical existence and consistency; it is a property of all relations, other than purely nominal relations, since it is in the minim and all relations possess the minim as a property. The second usage is **modal possibility**. This is the possibility that may be quantified as plural, singular, and zero, to give the three traditional **modalities** of **contingency**, **necessity**, and **impossibility**, respectively. The third usage is in modern **modal logic**, where the possible is defined as true in at least one possible world, and the necessary is defined as true in every possible world.

1. Relations and their Properties

Modern modal logic is part of a larger system of thought, called truth-functional logic, in which everything is defined in terms of truth and falsity (63, 231). We shall see in Chapter 3 that this has nominal meaning only, and as such is a travesty of reason; for now it is enough to show the incoherence of modal logic. What, in modal logic, is a possible world? To say that a possible world is a world that is true in at least one possible world is not only circular, it is incoherent. So we cannot know any such possibilities. And can we know all possible worlds? If not, then we cannot know any necessities. And if R necessitates S in one possible world because S is a singular possibility given R, but not in another possible world which does not contain R and S, then the statement that R necessitates S is not true in every possible world. So we may dismiss this purely nominal meaning of the word possibility, and hence that of nominal necessity.

The two other possibilities — logical and modal — are, surprisingly, identical (69). In the first place, impossibility — or zero modal possibility — has only nominal meaning and so no place in rational thought other than to discover, or state, errors. So we are only concerned with contingency and necessity: plural and singular possibility. There are also two kinds of existence: mathematical existence, which is the possibility in the minim (16); and **actual existence**, which is the existence of the empirical world and of the noumenal world (134). Every possible world — a plurality — has mathematical existence, so if mathematical existence is plural possibility — a contingency — and actual existence is necessity — a singular possibility — then logical possibility and modal possibility are identical. This is not as implausible as it first seems. That mathematical existence is contingent is shown by the fact that a possible world emerges from an axiom level; this emergence is necessary, but the axiom level itself is not — it is one of many possible axiom levels and this 'many possible' is a plural possibility, hence a contingency; so we talk of possible worlds, not necessary worlds. Similarly, in mathematics the choice of an axiom set is contingent but the theorems that are implied by it are necessary.

There are three advantages in this identification of logical possibility and modal possibility. The first is that it solves the traditional problem of being, which is the problem of how to define actual existence; the second is that it conforms to Occam's Razor; and the third is that it is often convenient to confirm logical necessity by showing it to be a singular possibility.

So the possibility in the minim is either plural possibility or singular possibility — contingency or necessity; if it is plural the relation possessing it has mathematical existence and if it is singular the relation possessing it has actual existence. We know of actual existence empirically: everything in the empirical world is actual. The identity of this actuality and of necessity — singular possibility — is far from obvious but will be shown in Chapter 8.

There is also **extrinsic possibility**: two or more relations are extrinsically possible, relative to each other, if they may be terms of one relation: each is a coterm of the other in that relation. That is, if R and S are terms of Q, so that R is a coterm of S in Q and S is a coterm of R in Q, then R and S are extrinsically possible. A broader definition is: two relations are extrinsically possible if they may both exist in one possible world; this follows transitively from the first definition. Extrinsic possibility is most important when it is singular, as in the necessity relations — demergence, emergence, inherence, causation, implication, and function — discussed in Section 3.

Probability is the reciprocal of modal possibility: of the three modalities, if a possibility has a degree $n > 1$, then it is a plural possibility and so a contingency, and $1/n=p$ is a probability that is neither a certainty nor an impossibility: $0 < p < 1$. If $n=1$ then we have a singular possibility and so a necessity; and $1/n$ is a probability of 1, or a necessity. And a possibility of $n < 1$ does not exist so a probability $p > 1$ does not exist and so *n* and *p* are then both impossible, and specifically so for *p* if *n =0*. This kind of probability is here called **modal probability** and distinguished from statistical probability — relative frequency — and the probability which is strength of belief in an explanation. The mathematics of probability leads to statistics, which is the mathematics of contingent sets.

We end this chapter with an introduction of some symbols. Please note that this book contains a glossary of defined symbols (247).

We represent a relation by an upper case letter, such as R, S, T, and a set by an upper case italic letter, such as *P, Q, R*. The terms of a relation R, which are themselves relations, are written to the right of R, as small caps, such as RAB, where upright A and B are particular terms of R, or as R*c* where italic *c* is the set of all the terms of R. The properties of R are written to the left of R, as small caps capped, such as P̂Q̂R, where P̂ and Q̂ are particular properties of R, or as *ŝ* R, where *ŝ* is the set of all the properties of R. The uppers of R are written as small cap superscripts of R, as R^S or R^s. In general, if *r̂* R^R*r* then the sets *r̂*, ^{R}, and *r* are defined as 'Every property of R', 'Every upper of R' and 'Every term of R'. However, when it is immaterial whether *p̂* is a single property or a set of properties, the upright form, P̂, will be used; similarly for R and *r*.

If R is a member of a set of relations *S* then this is written R∈*S*. Thus in a particular relation, R, if P̂RA and *ŝ* R*B* then P̂∈*ŝ* and A∈*B*.

Possibility is symbolised by ◊, intrinsic possibility by ◊̣, and extrinsic possibility by ◊̇; necessity by □, intrinsic necessity by ⊡, and extrinsic necessity by □̇.

We use the lower case letters p, q, r, etc. to represent statements of propositions; the usual ∧ for 'and' and ∨ for 'or' — the relational meaning of ∧ being the relation called a join (42); and a capital Greek letter such as Φ for any logical formula.

Because every intension defines an extension and every such defined extension has an intension, there are two functions relating intensions and extensions: the **relation-every**, symbolised by $\mathcal{E}$ and its inverse the **relation-any**, symbolised by $\mathcal{A}$. The usual symbolism for defining a set S intensionally is $S = \{x: x=p\}$ or $S = \{x| x=p\}$ which read: the set of every x such that x is a p; the braces, '{' and '}' indicate a set, the 'x:' or 'x|' read 'every x such that', where x is anything, and p is a statement of the intension of the set. A simplified symbolism is to drop the variable x and to use the relation-every: $\{\mathcal{E}p\}$; once again the braces indicate a set, $\mathcal{E}$ reads as 'every', and p is a statement of the intension. Since the relation-any is the inverse of the relation-every, $\mathcal{A}\{\mathcal{E}p\}$ is the intension p.

A second use of the symbol $\mathcal{E}$ is in relational quantificational logic: $(\mathcal{E}R)$, $(\mathcal{E}R)$, $(\mathcal{E}\hat{R})$, $(\mathcal{E}X)$, and $(\mathcal{E}x)$ are universal quantifiers, which are the use of the relation-every over specific universes of discourse; they differ from the first use of $\mathcal{E}$ both in having parentheses in place of braces and in having variables in place of intensions; and they differ symbolically from truth-functional quantificational logic (see Appendix B) in their variety of universes of discourse. We find these quantifiers useful in defining logical connectives, in Chapter 4, and they are also useful in Chapter 9, but they are not essential in relational logic: they merely make explicit the relation-every and the relation-any, which are usually obvious. It turns out that existential quantifiers are also not really needed in relational logic because of **existential presupposition** (56): this is the claim that whatever is being discussed exists. The rules of instantiation and generalisation of relational quantificational logic are the same as the rules in truth-functional quantificational logic. If existential quantifiers should be needed in relational logic, to emphasise existential presupposition, they could consist of the relation-any, as in $(\mathcal{A}R)$, $(\mathcal{A}\hat{R})$, $(\mathcal{A}\hat{r})$, $(\mathcal{A}X)$, and $(\mathcal{A}x)$. Note that $(\mathcal{A}R) \Leftrightarrow \Diamond R$, $(\mathcal{A}\hat{R}) \Leftrightarrow \Diamond\hat{R}$, and so on. In later chapters $\Diamond$ is preferred to $\mathcal{A}$.

1. Relations and their Properties

We use the symbol $\Rightarrow$ for the property of necessity, $\Leftarrow$ for its inverse, and $\Leftrightarrow$ when it is symmetric. The upward necessitation of emergence is symbolised by ⇑, and the downward necessitation of demergence by ⇓. Loop necessity is shown by ⇓⇑. The inherence of the properties of a relation in that relation, and its purely grammatical inverse, exherence, are symbolised by ↑↓.

The lowest level at which a property, R̂, or of its relation, R, emerge — their emergent levels — is #R̂ and #R, or #R̂R.

Dissimilarity and similarity are symbolised respectively by ≀ and ≈. These are derived from the symbol ~, for not, or negation. It is usual in logic that a proposition p is regarded as true unless it is written as ~p, in which case p is false; but sometimes it is necessary to emphasise that a proposition p is true: this is symbolised by ≀p; the symbol ≀ is the negation symbol, ~, rotated through a right angle, so that negation of ≀p is ~p and double negation is affirmation. We shall see (52) that one meaning of truth and falsity is similarity and dissimilarity, hence the symbols ≀ and ≈.

A stroke, /, through a symbol means its denial; p$\not\Rightarrow$q means ~(p$\Rightarrow$q).

Chapter 2. Relationally Defined Sets.

This chapter may be safely omitted on a first reading and by those uninterested in the fine detail of set theory. The object of this chapter is to show that all the fundamental concepts of set theory may be defined by means of relations, and so everything defined set-theoretically may be defined relationally.

So we are seeking definitions, in terms of relations, of the fundamental concepts in set theory: *intension, member, every, set, extension,* and *set-membership*. Note that a definition is itself a relation, having a *definiendum* and a *definiens* as terms.

An **intension** is a relation, R, and one term, T, of that relation, such as RT: a term of a relation and an upper of it.

A **member**, defined by an intension RT, is any coterm of T in R.

The **relation-every** is a compounded join (42) having the property we call completeness. Its terms are either the members defined by an intension or the denotations of an enumeration (50). It is symbolised by $\mathcal{E}$.

An **intensional set** is the compounded join of every member defined by an intension; as such it is a relation, called a **set relation**. It is symbolised by braces: {}. Thus an intensional set defined by an intension RT is[4] $\{\mathcal{E}RT\}$.

Set-membership is the demergence of a term of a set relation. Any such term is a member of the set that is the set relation. It is symbolised by $\in$.

----

[4] The standard usages for this are $\{x| \, x \in RT\}$ or $\{x: x \in RT\}$, where 'x|' or 'x:' reads 'every x such that', p is an intension, x is a variable, and $\in$ means set-membership; but $\{\mathcal{E}RT\}$, although synonymous with this, is preferable in the present context because it makes explicit the relation-every, $\mathcal{E}$. We could combine both notations, as in $\{\mathcal{E}x \ni x \in RT\}$, where '$\ni$' is the mathematicians' symbol for 'such that', but this seems unnecessary.

2. Relationally Defined Sets

The **extension** of an intensional set is the term set of the set-relation — the relation which is the compounded join which is the intensional set. Note that the set does not include the intension as a member; thus if R is demergence then $\{\mathcal{E}\text{R}\text{T}\}$ is the term set of T, which does not include T.

Intensions may be any of various relations or properties: relations of similarity or dissimilarity; necessities of exherence, demergence, or emergence; or boundaries. These six kinds of intension require that there be six kinds of intensional set; there may well be more kinds than this.

Examples of all of these definitions are:

A **similarity-intension** is a similarity and one of its terms R, say: $\aleph$R. This defines a similarity-set: $\{\mathcal{E}\aleph\text{R}\}$ or, more fully, $\{\mathcal{E}\text{X}\aleph\text{R}\}$, where X is a variable. This set is the set of every relation that is similar to R. A similarity set is an exception to the rule that T is not member of $\{\mathcal{E}\text{R}\text{T}\}$; this is because similarity, $\aleph$, is such that any member of $\{\mathcal{E}\aleph\text{R}\}$ may act as the intension of $\{\mathcal{E}\aleph\text{R}\}$; if one does so then R becomes a member of the set.

A **dissimilarity-intension** is a dissimilarity and one of its terms, R, say: $\approx$R. This defines a dissimilarity-set: $\{\mathcal{E}\approx\text{R}\}$ or, more fully, $\{\mathcal{E}\text{X}\approx\text{R}\}$. This set is the set of every relation that is dissimilar to R. $\{\mathcal{E}\text{X}\approx\text{R}\}$ is then the complement of $\{\mathcal{E}\text{X}\aleph\text{R}\}$, or $\{\mathcal{E}\text{X}\aleph\text{R}\}'$, the set of every non-R.

Similarity-intensions and dissimilarity-intensions are called **kind-intensions**: they define by kind of relation. All other intensions are **instance-intensions**: they define by instances of relations[5].

An **exherence-intension** is an exherence, E, and the term, R, which exheres the properties, $\hat{r}$, of R: $\uparrow\downarrow$ R. It defines

---

[5] Kind-intensions could equivalently be defined by a property set such as $\hat{r}$, the property set of R. Such a set would be $\{\mathcal{E}\aleph\hat{r}\}$ and $\{\mathcal{E}\hat{r}\aleph\hat{r}\} \supseteq \{\mathcal{E}\text{X}\aleph\text{R}\}$: the range of property sets is greater than the range of sets of relations because the former contains sets of properties which do not contain the minim.

the property set of R, $\{\mathcal{E}\hat{x}\uparrow\downarrow R\}$, hence $\{\mathcal{E}\uparrow\downarrow R\}=\hat{r}$. (Remember that $\hat{r}$ (italic case) is the set of all the properties of R while $\hat{x}$ (upright case) is a single property.)

A **demergence-intension** is a demergence, D, and its term, R: DR. It defines the term set of R — all the terms that demerge from R: $\{\mathcal{E}X\Downarrow R\}$ or $\{\mathcal{E}\Downarrow R\}$. This is the term set of R.

An **emergence-intension** is an emergence, $\Uparrow$, and a term of $\Uparrow$ which is an emergence configuration, C, of the term set of R, where *P* is the property set of R, such that *P* exheres R, as in the minor loop (20, 22) $R\Downarrow T\Rightarrow C\Uparrow P\Rightarrow R$.

A **boundary-intension** is an intension that defines a set by means of a boundary; it is a relation B such as *in*, *on*, or *bounded by*, which has the boundary as one of its terms and any possible relation within the boundary as its other terms. The set defined by a boundary-intension is called a **boundary set**. Examples of boundary sets are 'Everything in this box', 'All the people in this room', 'All the food on the table' and 'All the numbers between 1 and 100'.

We next define an **enumeration**, E, as a compounded join of names, and a **contingent set** as a set which does not have an intension, but is defined by a compounded join of the denotations of an enumeration. As such a contingent set is a relation, a set relation, defined as every denotation of E; its members are the terms of the set. With the exception of sets defined by kind-intensions, intensional sets may not be defined by extension. We see an example of this with the set of all living people, a set whose members are bounded by being between two times: birth and death; this set can in principle be enumerated, but the set of all people, which includes all future people, cannot be enumerated, even in principle.

An **extensional set** is either an intensional set or a contingent set. It is so called because it has extensional meaning. A **nominal set** is a name or description of an extensional set or of a null set. It is so called because it has nominal meaning.

Chapter 3. Relational Logic.

In this chapter we examine a hierarchy of five logics: relational logic, which consists of mathematical logic and intensional logic; extensional logic; exclusively extensional logic; and nominal logic. Mathematical logic is the logic used by mathematicians; intensional logic is the logic of intensions and their extensions; extensional logic is the logic of extensional sets, i.e. sets defined either by intension or by enumeration; exclusively extensional logic excludes the logic of intensional sets; and nominal logic is truth-functional logic and the logic of nominal meanings. In so far as relations are concerned, relational logic is primarily concerned with *kinds* and extensional logic with *instances*. Of these logics we can say that R⇒E⇒N. In this chapter we are mainly concerned with relational logic. Extensional logic is, of course, the logic of sets, which has ideal existence only; but we need to examine it and use it because we are born to classify (189).

The most important of all of these logics is mathematical logic, the logic used by mathematicians, which includes rules of inference not contained in any other logical system, such as mathematical induction and the rule of permissible modifications of equations — equals operated on by equals are equals — as well as relations such as equality — the basis of algebraic, differential, and other equations — and necessity relations such as functions, which other logics do not include. Mathematicians do not usually formalise this logic, probably because to them the necessities in it are obvious and do not need to be stated. Logicians, on the other hand, do not usually include mathematical argument forms in their logic because they are concerned with arguments in ordinary language, not those peculiar to mathematics.

Intensional logic is the logic of intensions and their extensions. Mathematical logic and intensional logic together constitute relational logic.

Apart from relational words or symbols used to record or communicate it, relational logic consists entirely of relations. It begins with abstract ideas, which are relations in the mind. A concept is a word or symbol bonded to an abstract idea; bonding is a relation. A proposition is either a structure of abstract ideas or a structure of concepts, and a structure is relations relating relations, to any number of levels. Propositions may include logical connectives and quantifiers, all of which are relations. A proposition is either true or false: either analytically (the predicate is, or is not, contained in subject) or synthetically (the proposition is either similar or dissimilar to reality); thus truth and falsity are relations. An argument is a structure of propositions and is valid if the truth of the premises necessitates the truth of the conclusion, or if the falsity of the conclusion necessitates the falsity of at least one of the premises; thus validity is a relation. Ancillary to all this are definitions and equivalences, which are relations. A suitable structure of all of these is a proof, which is a relation.

We next justify these claims.

It is unfortunate that relational logic and relational set theory are so interrelated that to some extent each presupposes the other. So, unavoidably, some things in this chapter presuppose the next chapter. They are relations between sets, such as subset, $\subset$; intersection, $\cap$; union, $\cup$; and complement, $'$, or non-. Also, relations between property sets such as subintension, $\prec$; superintension, $\succ$; commonality, $\curlywedge$; and coupling, $\curlyvee$. We are here primarily concerned with relations between these two kinds of relations. In so far as all these are unfamiliar to you, you should skip over them on a first reading and then come back to them later, or else jump forwards and backwards between the two chapters.

We begin, as usual, with some definitions.

Relational synthetic truth is defined by means of similarity. Two relations are similar if each member of the property set of each relation is similar to a member of the property set of the other relation. This concept may be extended to structures and wholes, as in the examples of congruence, matrix equality, and isomorphism. If a pair of relations, or structures, or wholes are not similar then they are dissimilar. If a relation, structure, or whole is a copy, representation, **image**, facsimile, or reproduction of another, and the two are similar, then their similarity is called the **relational synthetic truth**, **similarity truth**, or **correspondence truth** of the copy, relative to the other, or original. If they are not similar then the copy is **relationally false**, or **dissimilarity false**, relative to the original. If a kind of an ideal relation, $\hat{R}_2$, is similar to the kind, $\hat{R}_1$, of a noumenal relation or of an empirical relation, then the ideal kind is relationally synthetically true: $(\hat{R}_2 \wr\wr \hat{R}_1) \Leftrightarrow \wr \hat{R}_2$, where the symbol $\wr$ means true. Thus relational synthetic truth may be empirical or noumenal. And what is true of kinds is also true of instances, so that if an ideal relation R_2 is similar to a noumenal relation R_1 then the ideal R_2 is relationally synthetically true: $(R_2 \wr\wr R_1) \Leftrightarrow \wr R_2$. Equally, we can say that if $\hat{R}_2$ is dissimilar to $\hat{R}_1$, $(\hat{R}_2 \approx \hat{R}_1) \Leftrightarrow \sim \hat{R}_2$, then $\hat{R}_2$ is dissimilarity false; a special case of this occurs when $\hat{R}_1$ does not exist, as with the electromagnetic ether.

In the same way, since a relational **proposition** is a whole composed of ideal relations — abstract ideas — then if a relational proposition is similar to a portion of the noumenal world or of the empirical world, then the proposition is relationally synthetically true: it is usually a true portion of applied mathematics; and a statement of the proposition is the nominal meaning of the proposition and the statement is nominally true if the proposition is true. Similarly, a relational proposition is relationally and nominally false if it is dissimilar to reality. We are of course familiar with all this in everyday language, when we say of a portrait that it is, or is not, a good likeness of the person portrayed.

Furthermore, because we have degrees of similarity and dissimilarity (39, 182), there are degrees of truth and falsity.

So relational synthetic truth and falsity are relative to reality. **Empirical reality** is all that exists in the empirical

world that is not illusory, the empirical world being all that
exists and is perceptible, all that is known through the senses.
Noumenal reality is all that exists in the noumenal world, the
noumenal world being all existence that is underlying the
empirical world, all existence that is imperceptible. The two of
these together are the whole of reality and the basis of
synthetic truth: that is, reality is the original of which
propositions are an image, similar or dissimilar to that reality,
true or false. The search for empirical reality is empirical
science and the search for noumenal reality is theoretical
science; the latter succeeds in so far as it explains the former
— see Chapter 7. It is universally accepted by rational people
that all reality is logical, in the sense that it does not contain
contradictions. And noumenal reality is also logical in that it
contains necessities, such as causations — unlike empirical
reality, which has no necessities.

A second kind of relational falsity is relational
complementarity, discussed in the next chapter (75). If $\hat{s}$
defines a kind of relation then the complement of $\hat{s}$ is non-$\hat{s}$,
or $\hat{s}'$, so $\wr\hat{s}\,R_1 \Leftrightarrow \sim(\hat{s}'\,R_2)$ — the subscripts indicating that R_1
and R_2 are particular individual relations, particular instances
of R.

Relational analytic truth, as opposed to synthetic
truth, is based on necessity. At this point we need to fast
forward to the next chapter, where superintension (72) is
defined as $\hat{s} \succ \hat{p}$, where $\hat{s}$ and $\hat{p}$ are sets of properties of
relations and $\hat{p}$ is a subset of $\hat{s}$. If $\hat{s} \succ \hat{p}$ then it is analytically
true that $\hat{s}$ entails $\hat{p}$, $\hat{s} \Rightarrow \hat{p}$; and if $\hat{s} \wr\wr \hat{p}$ then it is analytically
true that $\hat{s}$ and $\hat{p}$ are relationally equivalent, $\hat{s} \Leftrightarrow \hat{p}$. The
symbols $\hat{s}$ and $\hat{p}$ are chosen in recognition of the fact that the
traditional definition of analyticity was that the predicate, $\hat{p}$, is
contained in the subject, $\hat{s}$; but note that the alternative
traditional definition of analyticity — that denial of an analytic
truth produces a contradiction — has no relational meaning
because neither does the word contradiction.

3. Relational Logic

A necessary condition for both relational analytic truth and relational synthetic truth is consistency, which is one of the properties of the minim — also known as possibility and as logical or mathematical existence. This will be called **precursor relational truth** and symbolised by $\diamond$. (Possibility is usually symbolised by $\Diamond$, but here we need to distinguish intrinsic possibility from extrinsic possibility (44) and we do so by internal and external dots: extrinsic possibility is symbolised by $\dot{\Diamond}$.) It is obvious that if either $\hat{P}$ or $\hat{Q}$ does not exist, then $\hat{S} \Rightarrow \hat{P}$ cannot be true. Thus the absence of precursor relational truth is **precursor relational falsity**.

We symbolise these three kinds of relational truth by the symbol $\wr$ and the three kinds of relational falsity by $\sim$. We use the symbol $\wr$ when it is needed for emphasis, as in $\wr$p, but otherwise use the convention of omitting it: 'p' normally means that p is true, while '$\sim$p' means that p is false. The symbol $\wr$ is chosen because the usual symbol for falsity is $\sim$; negation, or denial, rotates the symbols $\wr$ and $\sim$ through a right angle; so double negation is affirmation. (But, in accordance with English usage, $\wr\wr$p does not mean $\sim$p — although it would mean it if English were more logical.)

The **abstract idea** of a relation is an actual instance of that relation, in the mind, and **abstract thought** is manipulation of abstract ideas and discovery of their abstract properties, and of relations — other abstract ideas — between them. In particular, necessities in thought lead to thought about logic, mathematics, and explanation. Such thought, without the aid of language, is **pure thought**.

Pure thought is greatly aided by representing these mental relations, or their properties, by symbols or words: spoken, on paper, or in the mind. A symbol or word bonded to an abstract idea is then a **concept**. Such thought with concepts is what the nominalists want to call "silent speech", but, unlike the nominalist view, it does include abstract ideas. (In nominalism (65) all thought (which is abstract, as opposed to concrete imagination) is nothing but silent speech, and the meanings of the words in this silent speech are the words themselves, because there are no abstract ideas.) These concepts might be called **abstract concepts**, to distinguish them from **concrete concepts**, which are words bonded to concrete images; but we are not much concerned with concrete concepts in this book, and will confine the meaning of the word concept to the abstract. It must be emphasised that it is quite possible to think with abstract ideas alone — pure thought — as opposed to **conceptual thought**, which is thought with concepts, and **nominal thought**, which is thought with words alone — words without bonded abstract ideas or concrete images, as in algorithmic thought and thought that is no more than manipulation of words or symbols. Thought in ordinary language is conceptual thought adulterated with words having nominal meaning only: names of confused images or having vague denotation, and thought which is merely algorithmic, or else having no reference. But nonetheless conceptual thought is essential for communication of pure thought, and is not very difficult to separate from the thought of ordinary language. For the sake of completeness we can also define **extensional thought**, which is thought with extensions; for example, we might think of the family of Jack and Jill as the two of them plus all of their children, Jane, Jeff, Jill, Joan, John, and June. Such thought only operates with small extensions. Also, none of these different kinds of thought should be confused with imagining, which works with concrete images in the imagination.

Existential presupposition is the claim that whatever is being discussed exists. This applies to abstract thought because to think abstractly is to have abstract ideas — relations in the mind — which thereby exist; so relational meaning has at least ideal existence, and maybe real (empirical and/or noumenal) existence as well. The term existential presupposition was used in the past to distinguish traditional logic (229), which has it, from modern logic, which does not have it. In the present context it is precursor relational truth.

Thought that is no more than silent speech — purely nominal thought — does not, of course, have existential presupposition.

Relational validity is necessary transmission of truth or falsity. Relational analytic truth is superintension, as in $\hat{s} \succ \hat{P}$, so that $\wr\hat{s} \Rightarrow \wr\hat{P}$ and $\sim\hat{P} \Rightarrow \sim\hat{s}$, and the symbol $\Rightarrow$ represents implication, a necessity. The analytic truth provides the necessity and the transmitted truth or falsity is either synthetic truth or falsity; or else complementarity in the case of falsity. Upward necessity, ⇑, and downward necessity, ⇓, (47) also are analytically true and transmit truth and falsity, necessarily. **Valid argument forms** are forms of logical argument that are accepted as valid because their necessity is obvious, and which may be quoted in a formal argument; they usually contain $\Rightarrow$, and sometimes ⇑ or ⇓, which justifies their validity. **Valid equivalences** are symmetric argument forms; they are expressions containing the binecessity, $\Leftrightarrow$, either side of which may be substituted for the other in a logical argument. Similarities, such as $\hat{s} \wr\wr \hat{P}$, are also a basis for equivalence, since $(\hat{s} \wr\wr \hat{P}) \Leftrightarrow [(\wr\hat{s} \Leftrightarrow \wr\hat{P}) \vee (\sim\hat{s} \Leftrightarrow \sim\hat{P})]$.

The two great merits of modern symbolic logic, both adopted in all the following logics, are its use of symbols, in mathematical fashion, and its clear layout of a formal argument. This layout consists of a series of logical statements, each of which is on a new line, is numbered for purposes of cross-reference, and is logically justified on the grounds of being a premise or an axiom, a valid argument form or equivalence, a definition, or an obvious truth such as a tautology. Examples of this layout are on page 65, and in Chapter 9 and Appendix B.

We turn now to relational logic, which comprises mathematical logic and intensional logic. It is so called because it has relational meaning.

Mathematical logic has relations or properties of relations as its subject matter, such as *necessity, similarity,* *dissimilarity*, and *equality*. Necessity is known introspectively; implication and validity are necessary transmission, of truth, falsity, existence, or of concepts. Truth is similarity of an abstract idea to a part of reality, and thereby affirmation of the existence of that part of reality, or else superintension (72). So all valid inferences in mathematical logic are either confirmed introspectively or else validly derived from previously confirmed inferences.

Public mathematical logic is consensus on what is necessary.

Known connectives include:

> *Implication*, which is any relation having the relational property of necessity.
>
> *Conjunction*, which is either the relation of coupling (73) or of intersection (74).
>
> *Disjunction*, which is either the relation of commonality (73) or of union (73).
>
> *Equivalence*, which is either similarity or binecessity (60).
>
> *Equality*, a special case of equivalence (57, 58).
>
> *Existence*, which is intrinsic possibility.
>
> *Truth*, which is variously similarity, superintension, or affirmation (assertion of truth).
>
> *Negation* or *non-existence*, which is dissimilarity (complementarity) or else denial (absence of truth).

Known valid inferences include:

> *Modus ponens*.
>
> *Modus tollens*.
>
> *Disjunctive syllogism*.
>
> *Substitution of equivalents*.
>
> *Reductio* (denial only, of truth or of existence).
>
> *Conditional proof*.

Quantificational rules of instantiation and
generalisation.
Equations: equals operating on equals produce
equals.
Mathematical induction.
Known tautologies include: p⇒p, p∨~p, and ~(p∧~p)
— which are Aristotle's laws of thought.
Known equivalences include: ~~p⇔p,
~(p∧q)⇔(~p∨~q), and *dualities* (see
Appendix E).

Intensional logic, the other part of relational logic,
deals with intensions and their extensions.

With two limitations, all the standard argument forms
and equivalences of modern logic, including instantiation and
generalisation of the universal quantifier (the relation-every)
and its inverse (the relation-any), are obviously intensionally
valid; the more commonly used ones are:

Modus ponens: $[(\hat{s} \succ \hat{P}) \wedge \wr \hat{s}] \Rightarrow \wr \hat{P}.$
Modus tollens: $[(\hat{s} \succ \hat{P}) \wedge \sim \hat{P}] \Rightarrow \sim \hat{s}.$
Hyp. syllogism: $[(\hat{s} \succ \hat{P}) \wedge (\hat{P} \succ \hat{R})] \Rightarrow (\hat{s} \succ \hat{R}).$
Simplification: $\wr(\hat{s} \curlyvee \hat{P}) \Rightarrow \wr \hat{s}.$
Conjunction: $(\wr \hat{s} \curlyvee \wr \hat{P}) \Rightarrow \wr(\hat{s} \curlyvee \hat{P}).$
Disj. syllogism: $[\wr(\hat{s} \curlywedge \hat{P}) \wedge \sim \hat{P})] \Rightarrow \wr \hat{s},$
$[\wr(\hat{s} \curlywedge \hat{P}) \wedge \sim \hat{s})] \Rightarrow \wr \hat{P}.$
Contraposition: $(\hat{s} \succ \hat{P}) \Leftrightarrow (\sim \hat{P} \succ \sim \hat{s}),$
$(\hat{s} \succ \hat{P}) \Leftrightarrow (\hat{P}' \succ \hat{s}').$
Double negation: $\sim \sim \hat{s} \Leftrightarrow \wr \hat{s}, \sim \hat{s}' \Leftrightarrow \wr \hat{s}.$
DeMorgan: $(\hat{s} \curlyvee \hat{P}) \Leftrightarrow (\hat{s}' \curlywedge \hat{P}')',$
$(\hat{s} \curlywedge \hat{P}) \Leftrightarrow (\hat{s}' \curlyvee \hat{P}')',$
$(\hat{s} \curlyvee \hat{P}) \Leftrightarrow \sim(\sim \hat{s} \curlywedge \sim \hat{P}),$
$(\hat{s} \curlyvee \hat{P}) \Leftrightarrow \sim(\sim \hat{s} \curlywedge \sim \hat{P}).$

The first of the above mentioned limitations is that a
relational disjunction (74) must be complete (76); this

limitation is a feature of relational disjunction, discussed in the next chapter, that extensional and nominal disjunction do not have; it applies to the argument form called Disjunctive Syllogism, in that the disjunction in this syllogism must be complete. The second limitation is that the argument form called Disjunctive Addition, although valid in intensional logic, is harmless (82). If you try to formulate Disjunctive Addition in intensional logic it comes out as $\wr\hat{s}\Rightarrow\wr(\hat{s}\wedge\hat{P})$, which is a triviality. This is important because, although this argument form is allowed in lesser logics, it is highly undesirable because it allows the introduction of unlimited irrelevance into an argument.

Intensional implication is superintension: $\hat{s}\succ\hat{P}$ means $\hat{s}\Rightarrow\hat{P}$.

Logical equivalences are similarities: if $\hat{s}\wr\!\wr\hat{P}$ then $\hat{s}$ may be substituted for $\hat{P}$, or *vice versa*, in any expression that contains one of them. Equally so for binecessity: $\hat{s}\Leftrightarrow\hat{P}$; for example, $\hat{s}$ is an equilateral triangle and $\hat{P}$ is an equiangular triangle.

Intensions produce extensions, so relational logic includes the logic of intensional sets; this is determined by the implication theorem (79): $(\hat{A}\succ\hat{B})\Leftrightarrow(\{\mathcal{E}\hat{A}\}\subset\{\mathcal{E}\hat{B}\})$. Since superintension is relational analyticity, subset must be **extensional analyticity** and **extensional analytic truth**. Thus an **extensional inference** from A to B is $A\subset B$ and both **extensional implication** and **extensional analytic truth** are the relation of subset.

Extensional synthetic truth is the same as intensional synthetic truth: similarity. We have to say that sets have some reference to noumenal and empirical existence (135) because a set has a number of members and such numbers cannot be denied. Thus, empirically, the set of everything in this box has a number which is empirically real, as common sense asserts; and noumenally the set of protons in an atomic nucleus has a number which is noumenally real, as any chemist will agree. But the sets themselves are ideal only.

3. Relational Logic

Extensional synthetic falsity is therefore dissimilarity.

Precursor extensional truth is set-membership, and **precursor extensional falsity** is non-membership, or absence of membership; this kind of truth and falsity comes from comparing the two expressions:

$$(\hat{S} \succ \hat{P}) \Leftrightarrow [(\wr\hat{S} \Rightarrow \wr\hat{P}) \vee (\sim\hat{P} \Rightarrow \sim\hat{S})] \text{ and}$$
$$(S \subset P) \Leftrightarrow \{[(x \in S) \Rightarrow (x \in P)] \vee [(x \notin P) \Rightarrow (x \notin S)]\}.$$

These expression are derivative from the definitions of superintension and subset (72). This truth, set-membership, $\in$, may also be symbolised by $\wr$, and this falsity, $\notin$, by $\sim$. **Extensional contradiction** is then the null set, ϕ, (30, 74) and **extensional validity** is extensionally necessary — i.e. universal (62) — transmission of extensional truth or falsity.

From all this we get a parallel set of argument forms belonging to the logic of intensional sets, using p, q, and r as intensions:

Modus ponens: $[(\{\&p\} \subset \{\&q\}) \wedge (x \in \{\&p\})] \Rightarrow (x \in \{\&q\}).$
Modus tollens: $[(\{\&p\} \subset \{\&q\}) \wedge (x \notin \{\&q\})] \Rightarrow (x \notin \{\&p\}).$
Hyp. syllogism: $[(\{\&p\} \subset \{\&q\}) \wedge (\{\&q\} \subset \{\&r\})] \Rightarrow$
$(\{\&p\} \subset \{\&r\}).$
Simplification: $(x \in \{\&p\} \cap \{\&q\}) \Rightarrow (x \in \{\&p\}).$
Conjunction: $[(x \in \{\&q\}) \wedge (x \in \{\&p\})] \Rightarrow (x \in \{\&q\} \cap \{\&p\}).$
Disj. syllogism: $[x \in (\{\&p\} \cup \{\&q\}) \wedge (x \notin \{\&q\})] \Rightarrow$
$(x \in \{\&p\}).$
Disj. Addition: $\{\&p\} \subset \{\&p\} \cup \{\&q\}$
Contraposition: $(\{\&p\} \subset \{\&q\}) \Leftrightarrow (\{\&q\}' \subset \{\&p\}').$
Double Negation: $\sim(x \notin \{\&p\}) \Leftrightarrow (x \in \{\&p\}).$
DeMorgan: $[(x \in \{\&q\}) \wedge (x \in \{\&p\})] \Leftrightarrow$
$\sim[(x \notin \{\&q\}) \vee (x \notin \{\&p\})].$

The two above limitations to relational logic also apply here: a relational disjunction must be complete, and disjunctive addition, $\{\&\hat{P}\} \subset \{\&\hat{P}\} \cup \{\&\hat{Q}\}$, is relationally valid but only if $\{\mathcal{E}\hat{P}\}\cup\{\mathcal{E}\hat{Q}\}$ is an intensional set; see Theorem 4.10 (80). Logical equivalences also apply because of the equivalence theorem (79): $(\hat{A}\,\mathfrak{V}\,\hat{B})\Leftrightarrow(\{\mathcal{E}\hat{A}\}=\{\mathcal{E}\hat{B}\})$; if $\{\mathcal{E}\hat{A}\}=\{\mathcal{E}\hat{B}\}$, then $\hat{A}$ may be substituted for $\hat{B}$, or *vice versa*, in any expression in which they occur.

Extensional logic is the logic of extensional sets: a logic of sets which are defined either by intension or by enumeration, sets which may or may not have intensions; that is, extensional logic includes both necessary sets and contingent sets. Truth in this logic is extensional analytic truth, extensional synthetic truth, or precursor extensional truth, which is set membership. **Extensional necessity** is universality: the necessity of extensional implication is the universality of members of S being members of P in $S{\subset}P$. Some of its argument forms and equivalences are:

Modus ponens: $[(S{\subset}P)\wedge(x{\in}S)]\Rightarrow(x{\in}P)$.
Modus tollens: $[(S{\subset}P)\wedge(x{\notin}P)]\Rightarrow(x{\notin}S)$.
Hyp. syllogism: $[(S{\subset}P)\wedge(P{\subset}R)]\Rightarrow(S{\subset}R)$.
Simplification: $(x{\in}S{\cap}P)\Rightarrow(x{\in}S)$.
Substitution: $(S{=}P)\Leftrightarrow\{[(x{\in}S)\Leftrightarrow(x{\in}P)]\vee$
$[(x{\notin}P)\Leftrightarrow(x{\notin}S)]\}$.
Conjunction: $[(x{\in}P)\wedge(x{\in}S)]\Rightarrow(x{\in}P{\cap}S)$.
Disj. syllogism: $[(x{\in}S{\cup}P)\wedge(x{\notin}P)]\Rightarrow(x{\in}S)$.
Disj. addition: $(x{\in}S)\Rightarrow(x{\in}S{\cup}P)$.
Contraposition: $(S{\subset}P)\Leftrightarrow(P'{\subset}S')$.
Double Negation: ${\sim}(x{\notin}P)\Leftrightarrow(x{\in}P)$.
DeMorgan: $[(x{\in}Q)\wedge(x{\in}P)]\Leftrightarrow{\sim}[(x{\notin}Q)\vee(x{\notin}P)]$.

This logic allows unlimited Disjunctive Addition (as may be proved with a Venn diagram (see Appendix A)) and is not limited by incomplete disjunctions. And logical substitutions are allowed when $P{=}Q$.

Nominal logic is truth-functional logic (see Appendix B). In this, the same commonly used argument forms and equivalences are:

Modus ponens: $[(p \to q) \wedge p] \to q$
Modus tollens: $[(p \to q) \wedge \sim q] \to \sim p$
Hyp. syllogism: $[(p \to q) \wedge (q \to r)] \to (p \to r)$
Simplification: $(p \wedge q) \to p$
Conjunction: $(\text{ʔ}p \wedge \text{ʔ}q) \to \text{ʔ}(p \wedge q)$
Disj. syllogism: $[(p \vee q) \wedge \sim q] \to p$
Disj. addition: $p \to (p \vee q)$
Contraposition: $(p \to q) \equiv (\sim q \to \sim p)$
Double Negation: $\sim\sim p \equiv \text{ʔ}p$
DeMorgan: $(p \wedge q) \equiv \sim(\sim p \vee \sim q)$.

The important difference between nominal logic and all the other above logics is that this logic contains no genuine necessities. So called material implication, symbolised by $\to$, and equivalence, symbolised by $\equiv$, are defined exclusively by the truth or falsity of their terms: $(p \to q)$ is false if p is true and q is false, and is otherwise true; and $(p \equiv q)$ is true if p and q both have the same truth value, and is otherwise false. This truth is **nominal synthetic truth**, which is correct (i.e. established) usage of language, correct statement of meaning — relational, extensional, nominal, ideal, noumenal or empirical meaning — as opposed to **nominal synthetic falsity** which is incorrect usage, as in error or deceit. **Nominal analytic truth** or **nominal analyticity** is consistency, in the sense that it is not self-contradictory or does not lead to a contradiction, while **nominal analytic falsity** is self-contradiction. **Precursor nominal truth** is sense, as opposed to precursor nominal falsity which is nonsense. In truth-functional logic, truth and falsity are treated as primitive, undefined, and everything else defined by means of them. On this basis **truth-functional validity**, or **nominal validity** is simply tautology: always true. The difference between relational logic and truth-functional logic is that the necessities in the relational argument forms themselves necessitate — they are a sufficient condition for — the tautologies of the truth-functional argument forms, but the tautologies are only a necessary condition for the necessities. This gives truth-functional logic the appearance but not the reality of genuine logic. Notice that a tautology is at best an extensional necessity (62) — a universality, always true — and at worst purely nominal, as in "All mermaids are female".

The definitions of the five truth-functional connectives are given in Table 3.1, in their usual format — except that here the rows and columns are numbered for easy reference. This usual format conceals the fact that the rows and columns actually state propositions, and these propositions contain

	p	q	~p	p∧q	p∨q	p→q	p≡q
1	T	T	F	T	T	T	T
2	T	F	F	F	T	F	F
3	F	T	T	F	T	T	F
4	F	F	T	F	F	T	T
	1	*2*	*3*	*4*	*5*	*6*	*7*

Table 3.1.

conjunctions, ∧, implications, →, and negations, ~. For example, row-1 reads

$$(?p∧?q)→[~(~p)∧?(p∧q)∧?(p∨q)∧?(p→q)∧?(p≡q)]$$

and column-6 reads

$$[(?p∧?q)→?(p→q)]∧[(?p∧~q)→~(p→q)] ∧$$
$$[(~p∧?q)→?(p→q)]∧[(~p∧~q)→?(p→q)].$$

Also, part of column-3 reads

$$[⸮p→\sim(\sim p)]\wedge[(\sim p→⸮(\sim p)].$$

Clearly, the definitions of $\sim p$, $p\wedge q$, and of $p→q$ are useless because of circularity. (One could of course escape this circularity by replacing these three definitions with those of relational logic, but that would make the whole truth-functional approach pointless, since it is based exclusively on Boolean algebra (66).)

It is because of the inadequacies of truth-functional logic that it contains its well known paradoxes: a false statement validly implies any statement whatever and a true statement is validly implied by any statement whatever; and any two true statements are equivalent, as are any two false statements. The truth-functional proof that a contradiction proves q, anything whatever, is:

1. $⸮(p\wedge\sim p)$ Premise.
2. $\sim p$ 1, Simplification.
3. p   1, Simplification.
4. $p\vee q$ 3, Disjunctive addition.
5. q   4, 2, Disjunctive syllogism.

Truth-functional logic is endorsed primarily within a school of philosophy called **nominalism**, members of which — nominalists — deny the existence of abstract ideas. Three major such philosophers in English language philosophy were Bishop Berkeley (1685-1753), John Stuart Mill (1806-1873), and Ludwig Wittgenstein (1889-1951). Given that there are no abstract ideas it follows that, as opposed to imagination (25), all thought is silent speech, there is no thought without language, or, as Berkeley put it, words are the counters of the mind. Which is to say that everything abstract has nominal meaning only. Thus the meaning of an abstract word is the word itself, and such words are concrete. This is why truth-functional logic uses (concrete) statements rather than (abstract) propositions, denies necessity in implication, defines validity as tautology, calls relations polyadic predicates (237), and so on. If nominalists are right then there cannot be any synonyms: 'liquid' and 'fluid' have to have different meanings because they are different words; also, one word cannot have more than one meaning, so the metal *lead* is a dog's leash; and there cannot be translation from one language to another.

I suspect that nominalists are people with such vivid imaginations that they are blinded to the existence of abstract ideas, just as we are all blinded to the existence of stars during daylight. Nominalists are like some colour-blind people who refuse to believe that there are such things as colours: all talk of colours is, for them, an elaborate hoax.

It is important not to confuse truth-functional logic with the so-called truth-functional calculus. The latter is an algebraic structure of two-valued functions which is derived from undefined terms, such as 'propositions', p, q, r, etc., 'and', $\wedge$, and 'not', $\sim$; definitions from these, such as $(p \vee q) \equiv \sim(\sim p \wedge \sim q)$, and $(p \rightarrow q) \equiv \sim(p \wedge \sim q)$; axioms about these; and theorems derived from all this. The algebra is well known more generally as **Boolean algebra**, invented by George Boole (1815-1844). Based on mathematical logic, it is perfectly consistent and is used in gate theory in design of digital computers, in which truth and falsity are replaced with the binary numbers 1 and 0. In this algebra addition requires that $(1+1)=(1+0)=(0+1)=1$ and $(0+0)=0$; replacing 1 with 'true' and 0 with 'false' gives the truth-functional definition of disjunction. And Boolean multiplication requires that $(1 \times 0)=(0 \times 1)=(0 \times 0)=0$ and $(1 \times 1)=1$, which gives the truth-functional definition of conjunction. (This is why modern logicians refer to disjunction as a logical sum and to conjunction as a logical product.) Also, Boolean negation is addition of 1 modulo-2, so that $(1+1)=0$ and $(0+1)=1$: negation of truth is falsity and negation of falsity is truth. These three features of disjunction, conjunction and negation are the reason why modern logicians have identified Boolean algebra with logic. The identification is comparable with the claim that apples are oranges because they both have the three features of having a skin, of being spherical, and of being a fruit. The identification, in the logical case, fails because Boolean algebra cannot represent implication and equivalence.

For Boole, his search for an algebra of logic was like Columbus' search for a western route to India: each wrongly believed himself to have been successful, when in fact he had discovered something rather more valuable.

The hierarchy of all these logics is based on asymmetric necessity, $\Rightarrow$. This asymmetry is the basis of the distinction between sufficient and necessary conditions: If $A \Rightarrow B$ then A is a sufficient condition for B but not a necessary condition; and B is a necessary condition for A but not a sufficient condition; so (i) the presence of A requires that of B, while (ii) the absence of A does not require the absence of B; but (iii) the absence of B does require the absence of A and (iv) the presence of B does not require the presence of A. This applies not only to existence and non-existence hence also to possibility and impossibility, but also to truth and falsity. The three kinds of meaning also demonstrate this: $R \Rightarrow E \Rightarrow N$ means that the existence of R requires the existence of E which requires the existence of N; but the non-existence of N requires the non-existence of E which requires the non-existence of R — and this is the basis of *reductio ad absurdum*, or proof by contradiction, so that a nominal contradiction successively proves non-existence of N, E, and R, but cannot prove their existence. And similarly the truth of R requires the truth of E which requires the truth of N, but the falsity of N requires the falsity of E which requires the falsity of R — and these are the basis of *modus ponens* (affirmation of the antecedent) and *modus tollens* (denial of the consequent) — and also the fallacies of affirmation of the consequent (since truth cannot be proved from N to E to R), and of denial of the antecedent (since falsity cannot be proved from R to E to N). So it turns out that the fact of R⇒E⇒N means that there is a relational logic which justifies a mathematical logic and an intensional logic, and which also justifies an extensional logic which justifies an exclusively extensional logic, which justifies a nominal logic, and this justification is unilateral.

Chapter 4. Relational Set Theory.

In this chapter we examine relational set theory, and compare it to extensional and nominal set theories. Once again, we are including sets, which are ideal only, in our discussion because we are born to classify.

In logic, **connectives** are relations that combine concepts into other concepts or into propositions, or else propositions into larger propositions. In truth-functional logic they combine statements into larger statements.

Relational set theory is the theory of intensions and their connectives, extensional sets (29) and their connectives, and of the relations between these two kinds of connectives.

In contrast to the idea of similarity we next make precise the word **identity**. In everyday use 'identity' and 'similarity' are largely interchangeable — 'identity' meaning exact, or nearly exact, similarity as in 'identical twins'. In this book this usage is disallowed. Identity entails singularity and similarity entails plurality. If A and B are identical then they are one — also called one and the same, and numerically one — while if they are similar then they are two; in the first case A and B are two separate names of one thing, and in the second case they are separate names of two things. Paris and the capital of France are identical, *one*, while London and Paris are similar in being *two* great cities. Thus identity and similarity are mutually exclusive, disjoint, since self-similarity has nominal meaning only, because of endless multiplication.

To digress briefly, it may be remarked in this context that the everyday ambiguity of identity and similarity greatly aids the fallacy — here called the **identity error** — of inferring identity from similarity. A very common example in empiricist thought is identifying a hypothetical noumenal object with the evidence for it, so that the empirical status of the evidence supposedly makes the hypothetical object an empirical object: the object and the evidence are supposedly similar, and thereby identical, hence empirical. Another example is the belief that to see a photograph or television picture of a person is to see that person: something clearly false since you can converse with a person but not with their photograph. The identity error most often occurs with near similarities, where identity is impossible. This impossibility is because *any* qualitative difference entails quantitative difference, as is easily proved: whatever A and B may be, if there is some qualitative difference between them then there is some quality Q such that A is Q and B is not-Q, or *vice versa*; so if A and B are identical — numerically one — then one thing is at once Q and not-Q, which is impossible; therefore A and B are two. The identity error is masked in language by ambiguity with the words 'same' and 'different'. 'Same' may mean one and the same, identical, or else mean more or less similar; and 'different' may mean either qualitatively different or else quantitatively different. But here, similarity requires plurality and identity denies it.

In what follows the principle that qualitative difference entails quantitative difference will be abbreviated to **Q.Q.**

Similarity and identity are important in relational logic because of their connection to the distinction between *kind* and *instance* of relations: kinds are defined by similarity and instances by identity. If $\hat{r} R r$ and $\hat{s} S s$ are two instances of one kind of relation then $\hat{r}$ is similar to $\hat{s}$, $\hat{r} \wr \hat{s}$, but R and S are not identical, nor are r and s; but if R and S should be identical then so are $\hat{r}$ and $\hat{s}$ and so are r and s. Because of all this, *set-membership of kinds is defined by similarity* and *set-membership of instances is defined by identity*, as the following definitions and theorems will show. In general we are not interested in identity of relational properties, only in their similarities or dissimilarities. This means that in the case of kinds we only need to consider the property set, rather than instances of relations having this property set. So rather than symbolising the intension of a set as $\hat{r}R$, we usually simply use $\hat{r}$.

It may be that there are two kinds of set-membership, extensional and intensional, each symbolised by $\in$ and distinguished by context: $x \in \hat{s}$ is relational set-membership and $x \in S$ is extensional set membership; or maybe not. This will be left an open question because it is not important.

  We now define **relational connectives**, which relate kinds, by means of similarity; and **extensional connectives**, which relate instances, by means of identity. (The similarities and identities in these definitions are printed in italic font for emphasis of this distinction.) Once defined, we look at theorems relating these two kinds of connectives.

1.   Two property sets, or kinds of relation, $\hat{s}$ and $\hat{r}$, are **similar**, symbolised by $\hat{s} \mathcal{U} \hat{r}$, if each member of $\hat{s}$ is *similar* to a member of $\hat{r}$, and *vice versa*. In symbols this is:

$$(\hat{s} \mathcal{U} \hat{r}) \Leftrightarrow (\mathcal{E}\hat{x})(\mathcal{E}\hat{y})(\mathcal{E}\hat{z})\{[(\hat{x}\mathcal{U}\hat{y}) \wedge (\hat{y} \in \hat{s})] \Leftrightarrow [(\hat{x}\mathcal{U}\hat{z}) \wedge (\hat{z} \in \hat{r})]\}.$$

(Remember that $\hat{x}$ and $\hat{y}$ are single properties and $\hat{s}$ and $\hat{r}$ are sets of properties (45).)

  Similarity is **relational equivalence.**

2.   Two extensional sets, S and T, are **identical**, symbolised by $S = T$, if each member of S is *identical* with a member of T, and *vice versa:*

$$(S = T) \Leftrightarrow (\mathcal{E}x)[(x \in S) \Leftrightarrow (x \in T)]^6.$$

Note that what is usually mis-called **set equality** is here set identity: to say of one set that it is equal to itself is to assume a monadic relation of equality, which is pointless. But to say that S and T are identical is to say that each of the symbols S and T name the one identical set. Because of this the symbol '=' is

---

  [6] In this and the following definitions of connectives, the use of braces, {, }, and brackets, [,], as well as parentheses, (,), for delimitation, is for clarity; the braces do not define sets, nor the parentheses quantifiers, unless the context clearly indicates otherwise, as with ($\mathcal{E}$x).

unfortunate: the mathematical '≡' would be much more suitable; however, the use of '=' in this context is well established so will be continued here.

Set identity is **extensional equivalence**

Thus we have used similarity of individual properties, Ŷ and Ẑ, to define similarity of property sets, in terms of *kinds* of relations, and identity of members, x, to define set identity, in terms of *instances* of relations

We invoke Occam's Razor here and say that no two members of a property set of any one relation are similar.

3.   A property set, $\hat{s}$, is a **subintension** of another property set, $\hat{t}$, symbolised $\hat{s} \prec \hat{t}$, if each member of $\hat{s}$ is *similar* to a member of $\hat{t}$, but not *vice versa*. In symbols this is:

$$(\hat{s} \prec \hat{t}) \Leftrightarrow (\mathcal{E}\hat{x})\{[(\hat{x} \in \hat{s}) \Rightarrow (\hat{x} \in \hat{t})] \wedge [(\hat{x} \in \hat{t}) \not\Rightarrow (\hat{x} \in \hat{s})]\}.$$

The inverse of subintension is **superintension**, symbolised by $\succ$: $(\hat{t} \succ \hat{s}) \Leftrightarrow (\hat{s} \prec \hat{t})$. If $\hat{s}$ is either a subintension of $\hat{t}$ or is similar to $\hat{t}$, this is symbolised $\hat{s} \divideontimes \hat{t}$, and its inverse by $\hat{t} \divideontimes \hat{s}$; these are called **improper subintension** and **improper superintension**.

Superintension is the main form of **relational implication**.

4.   An extensional set, S is a **subset** of another extensional set, T, symbolised $S \subset T$, if each member of S is *identical* with a member of T, but not *vice versa*. In symbols this is:

$$(S \subset T) \Leftrightarrow (\mathcal{E}x)\{[(x \in S) \Rightarrow (x \in T)] \wedge [(x \in T) \not\Rightarrow (x \in S)]\}.$$

The inverse of subset is **superset**, symbolised by $\supset$. If S is either a subset of T, or identical with T, this is symbolised by $S \subseteq T$, and its inverse by $T \supseteq S$; these are **improper subset** and **improper superset**.

4. Relational Set Theory

Subset is **extensional implication**.

5.   The **coupling** of two property sets, $\hat{s}$ and $\hat{r}$, symbolised by $\hat{s} \curlyvee \hat{r}$, is such that $\hat{s} \curlyvee \hat{r}$ is an improper superintension both of $\hat{s}$ and of $\hat{r}$:

$$(\hat{s} \curlyvee \hat{r}) \Leftrightarrow \{[(\hat{s} \curlyvee \hat{r}) \maltese \hat{s}] \wedge [(\hat{s} \curlyvee \hat{r}) \maltese \hat{r}]\}.$$

An example of coupling is provided by the minim (16), $\hat{M}$: if $\hat{A}$, $\hat{S}$, and $\hat{P}$ represent the properties of adicity, simplicity, and possibility then $\hat{M} \wr\wr (\hat{A} \curlyvee \hat{S} \curlyvee \hat{P})$.

Coupling of property sets is **relational conjunction**.

A relational conjunction is a representative instance (89) of each of its conjuncts.

6.   The **union** of two extensional sets, S and T, symbolised by $S \cup T$, is such that $S \cup T$ is an improper superset both of S and of T:

$$(S \cup T) \Leftrightarrow \{[(S \cup T) \supseteq S] \wedge [(S \cup T) \supseteq T]\}.$$

Union of extensional sets is **extensional disjunction**.

7.   The **commonality** of two property sets, $\hat{s}$ and $\hat{r}$, symbolised by $\hat{s} \wedge \hat{r}$ is such that each member of $\hat{s} \wedge \hat{r}$ is an improper subintension of both $\hat{s}$ and $\hat{r}$:

$$(\hat{s} \wedge \hat{r}) \Leftrightarrow \{[(\hat{s} \wedge \hat{r}) \maltese \hat{s}] \wedge [(\hat{s} \wedge \hat{r}) \maltese \hat{r}]\}.$$

If the commonality of $\hat{s}$ and $\hat{r}$, other than the three universal properties of possibility, simplicity, and adicity, does not exist, $\hat{s}$ and $\hat{r}$ are said to be **disparate**. Thus two disparate relations have no properties in common other than the fact of being relations: they are possible, simple, and have an adicity; this minimal commonality of three properties being the minim, $\hat{M}$ (16).

Since the definition of commonality allows it to be polyadic we may speak of the commonality of all the members of an intensional set; if $\hat{s}$ is an intensional set then the commonality of all of its members will be symbolised by $\wedge\hat{s}$. Because no two members of the property set of any one relation are similar, the commonality of such a property set does not exist.

It is in fact useful to extend the definition of $\wedge\hat{s}$ to all extensional sets: $\wedge S$ then means that S is either an intensional set or a contingent set. The utility of $\wedge S$ is explained shortly (see Theorems 4.3 and 4.4 below).

Commonality of property sets is **relational disjunction**

8.   The **intersection** of two extensional sets, S and T, symbolised by $S\cap T$, if it exists, is such that each member of $S\cap T$ is an improper subset of both S and of T:

$$(S\cap T)\Leftrightarrow\{[(S\cap T)\subseteq S]\wedge[(S\cap T)\subseteq T]\}.$$

If the intersection of S and T does not exist, S and T are said to be **disjoint**. (Note: the null set, ϕ, is not an intensional set, it is only a nominal set (29).)

Intersection of extensional sets is **extensional conjunction**.

9.   The **decoupling** of two property sets, $\hat{s}$ and $\hat{r}$, which are not disparate, symbolised $\hat{s}$-$\hat{r}$, if it exists, is the set consisting of those members of $\hat{s}$ which are *not similar* to any member of $\hat{r}$.

$$(\mathcal{E}\hat{x})\{\hat{x}\in(\hat{s}\text{-}\hat{r})\Leftrightarrow[(\hat{x}\in\hat{s})\wedge\sim(\hat{x}\in\hat{r})]\}.$$

But if $\hat{s}$ and $\hat{r}$ are the property sets of any two relations then both $\hat{s}$ and $\hat{r}$ contain the minim; yet $\hat{s}$-$\hat{r}$ and $\hat{r}\succ\hat{M}$ together require that $\sim[(\hat{s}\text{-}\hat{r})\succ\hat{M}]$, in which case $\hat{s}$-$\hat{r}$ does not define a relation; equally with $\hat{r}$ - $\hat{s}$, given that $\hat{s} \succ \hat{M}$. So decoupling is significant in the relational set theory that deals with every possible property set, but not in the relational set theory that deals only with sets containing the minim — with sets of relations, in other words.

10.   The **set difference** of two intersecting extensional sets, *S* and *T*, symbolised *S-T*, is the set consisting of those members of *S* which are *not identical* with any member of *T*:

$$(\mathcal{E}x)\{[(x \in (S\text{-}T)] \Leftrightarrow [(x \in S) \wedge (x \notin T)]\}.$$

11.   We define a **similarity set** as $\{\mathcal{E}(\mathcal{U}\hat{s})\}$: the set of all property sets that are similar to $\hat{s}$; that is, if $\hat{r} \in \{\mathcal{E}\mathcal{U}\hat{s})\}$ then each member of $\hat{s}$ is similar to a member of $\hat{r}$, and *vice versa*, and $\hat{s}$ and $\hat{r}$ are in one-to-one correspondence. (See degree of similarity (39): $\hat{r}\mathcal{U}\hat{s}$ means that there is no degree of dissimilarity between them.) If $\hat{s} \succ \hat{M}$ and $\hat{s}$R then $\{\mathcal{E}(\mathcal{U}\hat{s}R)\}$ defines the set of all relations similar to $\hat{s}$R, which is to say all *instances* of $\hat{s}$R, and $\hat{s}$ defines the *kind* of relation R.

A similarity set could be said to define a class because it is the basis of classification.

We also define a **dissimilarity set** as $\{\mathcal{E}(\approx \hat{s})\}$: the set of all sets of properties that have at least one member dissimilar to every member of $\hat{s}$; that is, if $\hat{r} \in \{\mathcal{E}(\approx \hat{s})\}$ then at least one member of $\hat{r}$ is dissimilar to each member of $\hat{s}$, and/or *vice versa*. $\hat{s} \approx \hat{r}$ allows, but does not require, that every member of $\hat{s}$ may be dissimilar to every member of $\hat{r}$; and $\hat{s}$ and $\hat{r}$ may or may not be in one-to-one correspondence. So two relations R and S may be dissimilar even though they have some properties in common, such as the minim, $\hat{M}$.

Next, the **relational complement** of $\mathcal{U}\hat{s}$ is $\approx \hat{s}$, which may also be written $(\mathcal{U}\hat{s})'$ and which reads as $(\mathcal{U}\hat{s})$-prime, or as non-$\mathcal{U}\hat{s}$, and if $\hat{U}_{\hat{p}}$ is the universe of discourse of all properties then $[\hat{U}_{\hat{p}} - \{\mathcal{E}(\mathcal{U}\hat{s})]\mathcal{U}(\mathcal{E}(\approx \hat{s})\mathcal{U}(\mathcal{E}(\mathcal{U}\hat{s})'$.

12.   For the next definition of an extensional connective we assume the existence of an extensional set, U_R, called the universe of discourse, consisting of the set of every relation. If U_R is the universe of discourse and S is an extensional set within U_R then the **extensional complement** of S, symbolised S', is $S' = U_R - S$. S' is such that $S \cup S' = U_R$ and each member of S is *not identical* with each member of S', and *vice versa*. S' reads as S-prime, and also as non-S.

We now have three different universes of discourse. One is $U_{\hat{P}}$, the set of all properties. Another is U_R, the set of all relations: that is, all kinds of relations plus all instances of each kind. And third is $U_{\hat{P} \succ \hat{M}}$, the set of all kinds of relations. Clearly, $U_{\hat{P}} \cap U_R = U_{\hat{P} \succ \hat{M}}$. In what follows the relevant universe of discourse will usually be discernable from the context; if it is not, it will be stated explicitly.

Thus the relational complement is based on $U_{\hat{P}}$ and the extensional complement is based on U_R.

Finally, we introduce a peculiarity of relational disjunction.

A **complete disjunction** is a relational commonality, or disjunction, whose extension is an intensional set.

An **incomplete disjunction** is a relational commonality, or disjunction, whose extension is not an intensional set; that is, it is a contingent set, a set that does not have an intension.

Two examples will make this clear. Consider first the possibility of a natural number being either odd or even. If a natural number is Ń, odd is Ô, and even is Ê, then whatever is Ń is either (ŃⳐÔ) or (ŃⳐÊ); this is expressed relationally as (ŃⳐÔ)⅄(ŃⳐÊ), and [(ŃⳐÔ)⅄(ŃⳐÊ)]ꝶŃ. That is, anything that is either an odd number or an even number is a number; this is a complete disjunction.

For the second example, let the relational meanings of *simple plane polygon, trilateral*, and *quadrilateral* be P̂, T̂, and

Q̂. Then T̂γP̂ and Q̂γP̂ are the relational meanings of *triangle* and *quadrangle*, and {Ɛ(T̂γP̂)} and {Ɛ(Q̂γP̂)} are the intensional sets of every triangle and of every quadrangle. So {Ɛ(T̂γP̂)}∪{Ɛ(Q̂γP̂)} is the extensional disjunction of these disjoint sets; anything belonging to this set is, extensionally, either a triangle or a quadrangle. However the relational disjunction of T̂γP̂ and Q̂γP̂ is (T̂γP̂)ʎ(Q̂γP̂), and [(T̂γP̂)ʎ(Q̂γP̂)]ℜP̂: the commonality of T̂γP̂ and Q̂γP̂ is P̂. So anything which is a member of {Ɛ((T̂γP̂)ʎ(Q̂γP̂))} is a member of {ƐP̂}. Hence, relationally, anything which is either a triangle or a quadrilateral is, more accurately, either a triangle or a quadrilateral or any other simple plane polygon; that is, any simple plane polygon. In symbols this means:

$$\{Ɛ((T̂γP̂)ʎ(Q̂γP̂))\} ⊃ [\{Ɛ(T̂γP̂)\}∪\{Ɛ(Q̂γP̂)\}].$$

This relational "disjunction" is peculiar: it is one or the other or neither, where the *neither* refers to the set difference between the intensional set and the extensional set:

$$\{Ɛ((T̂γP̂)ʎ(Q̂γP̂))\} - [\{Ɛ(T̂γP̂)\}∪\{Ɛ(Q̂γP̂)\}].$$

This is the kind of relational disjunction that is incomplete: it is incomplete because the corresponding union is a contingent set: it produces an extensional disjunction, but not a relational one. Another illustration of incomplete disjunction is given later (83).

These two examples illustrate a criterion for complete and incomplete disjunction. {Ɛ(N̂γÔ)}∪{Ɛ(N̂γÊ)}={ƐN̂}, whereas [{Ɛ(T̂γP̂)}∪{Ɛ(Q̂γP̂)}]⊂{ƐP̂}, and {Ɛ(N̂γÔ)}∪{Ɛ(N̂γÊ)} is a partition[7] of {ƐN̂}, whereas [{Ɛ(T̂γP̂)}∪{Ɛ(Q̂γP̂)}] is not a partition of {ƐP̂}. So in general

---

[7] A partition of a set *S* is a set of subsets of *S* such that each subset is disjoint from every other subset in the partition, and the union of all of them is identical with *S*.

a relational disjunction $(\hat{A}\wedge\hat{B}\wedge...\hat{D})$ is complete if and only if $[\{\mathcal{E}\hat{A}\}\cup\{\mathcal{E}\hat{B}\}\cup...\{\mathcal{E}\hat{D}\}]$ is a partition of $\{\mathcal{E}(\hat{A}\wedge\hat{B}\wedge...\hat{D})\}$; otherwise the disjunction is incomplete. See Theorem 4.10 below.

All of the above definitions give rise to the following thirteen theorems relating the relational connectives to their corresponding extensional connectives. Because we are dealing with relations and their properties, and extensional sets (29), we have existential presupposition (56) and so no exclusively nominal sets are involved.

Theorem 4.1, the intension theorem: for any intension, $\hat{P}$, $\wedge\{\mathcal{E}\hat{P}\}\,ℛ\,\hat{P}$.

> *Proof.* $\{\mathcal{E}\hat{P}\}$ defines the set of every $\hat{P}$R, every relation of kind $\hat{P}$, which is $\{\mathcal{E}\hat{x}\ni(\hat{x}\,ℛ\,\hat{P})\}$. (Note: the symbol $\ni$ reads 'such that'.) So $(\mathcal{E}\hat{x})[(\hat{x}\in\{\mathcal{E}\hat{P}\})\Rightarrow(\hat{x}\,ℛ\,\hat{P})]$ $\therefore \wedge\{\mathcal{E}\hat{P}\}\,ℛ\,\hat{P}$.

Theorem 4.2, the extension theorem: for any extensional set S, $S\subseteq\{\mathcal{E}(\wedge S)\}$.

> *Proof.* $(\mathcal{E}\hat{x})(\hat{x}\in S)\Rightarrow(\hat{x}\,ℛ\,(\wedge S))\Rightarrow(\hat{x}\in\{\mathcal{E}(\wedge S)\})$, so $(x\in S)\Rightarrow(x\in\{\mathcal{E}(\wedge S)\})$. Therefore $S\subseteq\{\mathcal{E}\wedge S\}$, by definition of $\subseteq$.

Theorem 4.3, the intensional set theorem: An extensional set S is an intensional set if and only if $S=\{\mathcal{E}(\wedge S)\}$: $(S=\{\mathcal{E}(\wedge S)\})\Leftrightarrow(S$ is an intensional set$)$.

> *Proof.* (i) Suppose that $S=\{\mathcal{E}(\wedge S)\}$: then S is an intensional set, since $\{\mathcal{E}(\wedge S)\}$ is an intensional set. (ii) Suppose that S is an intensional set, $\{\mathcal{E}\hat{s}\}$; by Theorem 4.1, $\wedge\{\mathcal{E}\hat{s}\}\,ℛ\,\hat{s}$; therefore $S=\{\mathcal{E}\hat{s}\}=\{\mathcal{E}(\wedge\{\mathcal{E}S\})\}=\{\mathcal{E}(\wedge S)\}$, by the principle of substitution of equivalents.

Theorem 4.4, the contingent set theorem: An extensional set S is a contingent set if and only if $S \subset \{\mathcal{E}(\wedge S)\}$:

$S \subset \{\mathcal{E}(\wedge S)\} \Leftrightarrow (S$ is a contingent set$)$

> *Proof.* (i) Suppose that S is a contingent set; then S is not an intensional set, by definition of extensional set, in which case $S \neq \{\mathcal{E}(\wedge S)\}$, by Th. 4.3. So by Th. 4.2. $S \subset \{\mathcal{E}(\wedge S)\}$. (ii) Suppose that $S \subset \{\mathcal{E}(\wedge S)\}$; then S is not an intensional set, by Th. 4.3., in which case S is a contingent, by definition of extensional set.

Theorem 4.5, the equivalence theorem:

$(\hat{P} \, \wr \, \hat{Q}) \Leftrightarrow (\{\mathcal{E}\hat{P}\} = \{\mathcal{E}\hat{Q}\}).9*2$

> *Proof.* $[\{\mathcal{E}\hat{P}\} = \{\mathcal{E}\hat{Q}\}] \Leftrightarrow (\mathcal{E}\hat{x})[(\hat{x} \in \{\mathcal{E}\hat{P}\}) \Leftrightarrow (\hat{x} \in \{\mathcal{E}\hat{Q}\})]$ $\Leftrightarrow (\hat{P} \, \wr \, \hat{Q})$. (Reminder: $\hat{x}$ is a single property while $\hat{P}$ and $\hat{Q}$ are sets of properties.)

Corollary. $(\hat{P} \approx \hat{Q}) \Leftrightarrow (\{\mathcal{E}\hat{P}\} \neq \{\mathcal{E}\hat{Q}\})$.

Theorem 4.6, the implication theorem:

$(\hat{P} \succ \hat{Q}) \Leftrightarrow (\{\mathcal{E}\hat{P}\} \subset \{\mathcal{E}\hat{Q}\})$.

> *Proof.* By definition $(\hat{P} \succ \hat{Q}) \Leftrightarrow (\mathcal{E}\hat{x})\{[(\hat{x} \in \{\mathcal{E}\hat{Q}\}) \Rightarrow (\hat{x} \in \{\mathcal{E}\hat{P}\})] \wedge [(\hat{x} \in \hat{Q}) \not\Rightarrow (\hat{x} \in \hat{P})]\} \Leftrightarrow (\{\mathcal{E}\hat{P}\} \subset \{\mathcal{E}\hat{Q}\})$; hence $(\hat{P} \succ \hat{Q}) \Leftrightarrow (\{\mathcal{E}\hat{P}\} \subset \{\mathcal{E}\hat{Q}\})$.

Theorem 4.7, the conjunction theorem:

$\{\mathcal{E}(\hat{P} \curlyvee \hat{Q})\} = \{\mathcal{E}\hat{P}\} \cap \{\mathcal{E}\hat{Q}\}$.

> *Proof.* $(\hat{P} \curlyvee \hat{Q}) \succ \hat{P}$ and $(\hat{P} \curlyvee \hat{Q}) \succ \hat{Q}$ $\therefore (\hat{P} \curlyvee \hat{Q}) \succ (\hat{P}$ and $\hat{Q})$ $\therefore \{\mathcal{E}(\hat{P} \curlyvee \hat{Q})\} = \{\mathcal{E}\hat{P}\} \cap \{\mathcal{E}\hat{Q}\}$.

Theorem 4.8. For all intensional sets $A, B,$

$\wedge (A \cup B) \wr [(\wedge A) \curlywedge (\wedge B)]$.

> *Proof.* Suppose that $A = \{A_1, \ldots A_n\}$ and $B = \{B_1, \ldots B_m\}$; then $A \cup B = \{A_1, \ldots A_n, B_1, \ldots B_m\}$. $\therefore \ \wedge (A \cup B) \wr (A_1 \curlywedge \ldots A_n \curlywedge B_1 \curlywedge \ldots B_m) \wr [(\wedge A) \curlywedge (\wedge B)]$.

Theorem 4.9, the disjunction theorem:
$\{\mathcal{E}(\hat{P}\curlywedge\hat{Q})\}\supseteq\{\mathcal{E}\hat{P}\}\cup\{\mathcal{E}\hat{Q}\}$.

> *Proof.* $\curlywedge(\{\mathcal{E}\hat{P}\}\cup\{\mathcal{E}\hat{Q}\})\varkappa(\curlywedge\{\mathcal{E}\hat{P}\}\curlywedge\curlywedge\{\mathcal{E}\hat{Q}\})\varkappa(\hat{P}\curlywedge\hat{Q})$, by Theorems 4.8 and 4.1; but, by Theorem 4.2, $(\{\mathcal{E}\hat{P}\}\cup\{\mathcal{E}\hat{Q}\})\subseteq\{\mathcal{E}\curlywedge(\{\mathcal{E}\hat{P}\}\cup\{\mathcal{E}\hat{Q}\})\}$. So $\{\mathcal{E}(\hat{P}\curlywedge\hat{Q})\}\supseteq\{\mathcal{E}\hat{P}\}\cup\{\mathcal{E}\hat{Q}\}$.

Corollary. $\{\mathcal{E}(\hat{P}\curlywedge\hat{Q})\}=\{\mathcal{E}\hat{P}\}\cup\{\mathcal{E}\hat{Q}\}\Leftrightarrow$ (the disjunction is complete).

Theorem 4.10, the disjunctive addition theorem.
$\{\mathcal{E}\hat{P}\}\subset\{\mathcal{E}\hat{P}\}\cup\{\mathcal{E}\hat{Q}\}$ if and only if the disjunction is complete, which is true if and only if $\{\mathcal{E}\hat{P}\}\cup\{\mathcal{E}\hat{Q}\}$ is an intensional set. And the disjunction is incomplete if and only if $\{\mathcal{E}\hat{P}\}\cup\{\mathcal{E}\hat{Q}\}$ is a contingent set.

> *Proof.* By definition, a disjunction $\{\mathcal{E}\hat{P}\}\cup\{\mathcal{E}\hat{Q}\}$ is complete if and only if $\{\mathcal{E}\hat{P}\}\cup\{\mathcal{E}\hat{Q}\}$ is a partition of $\{\mathcal{E}(\hat{P}\curlywedge\hat{Q})\}$, in which case $\{\mathcal{E}\hat{P}\}\cup\{\mathcal{E}\hat{Q}\}=\{\mathcal{E}(\hat{P}\curlywedge\hat{Q})\}$, by definition of a partition. On the other hand, if $\{\mathcal{E}\hat{P}\}\cup\{\mathcal{E}\hat{Q}\}$ is not a partition of $\{\mathcal{E}(\hat{P}\curlywedge\hat{Q})\}$ then $\{\mathcal{E}\hat{P}\}\cup\{\mathcal{E}\hat{Q}\}\subset\{\mathcal{E}(\hat{P}\curlywedge\hat{Q})\}$, by Theorem 4.9 and by similar reasoning $\{\mathcal{E}\hat{P}\}\cup\{\mathcal{E}\hat{Q}\}$ is a contingent set. And if $\{\mathcal{E}\hat{P}\}\cup\{\mathcal{E}\hat{Q}\}$ is a contingent set then it is not a partition of $\{\mathcal{E}(\hat{P}\curlywedge\hat{Q})\}$.

Theorem 4.11, the negation theorem: $\{\mathcal{E}\hat{P}\}'=\{\mathcal{E}\hat{P}'\}$, $\{\mathcal{E}\hat{P}'\}'=\{\mathcal{E}\hat{P}\}$, and $\hat{P}\varkappa\hat{P}''$.

> *Proof.* $\hat{P}\curlywedge\hat{P}'$ is a complete disjunction so $\{\mathcal{E}\hat{P}\}\cup\{\mathcal{E}\hat{P}'\}$ is a partition of $\{\mathcal{E}(\hat{P}\curlywedge\hat{P}')\}$; and $(\hat{P}\curlywedge\hat{P}')\varkappa\hat{M}$. So $\{\mathcal{E}(\hat{P}\curlywedge\hat{P}')\}=\{\mathcal{E}\hat{M}\}=\hat{U}$, hence $\{\mathcal{E}\hat{P}\}\cup\{\mathcal{E}\hat{P}'\}$ is a partition of $\hat{U}$; but $\{\mathcal{E}\hat{P}\}\cup\{\mathcal{E}\hat{P}\}'$ is also a partition of $\hat{U}$; so $\{\mathcal{E}\hat{P}'\}=\{\mathcal{E}\hat{P}\}'$. Thus $\{\mathcal{E}\hat{P}'\}'=\{\mathcal{E}\hat{P}\}''=\{\mathcal{E}\hat{P}\}$. And by Theorem 4.5, $\hat{P}\varkappa(\hat{P}'')$.

These first eleven theorems show that there are duals (244) between relational connectives and extensional connectives. With two *caveats* they are:

$$\curlywedge \text{ and } \cup$$
$$\curlyvee \text{ and } \cap$$
$$\succ \text{ and } \subset$$
$$\prec \text{ and } \supset$$
$$\divideontimes \text{ and } \subseteq$$
$$\divideontimes \text{ and } \supseteq$$
$$\natural \text{ and } =$$

The two *caveats* are that (i) the extensional connectives must relate intensional sets, and (ii) all disjunctions must be complete.

Theorems 4.12 and 4.13, the distribution theorems:
$\hat{P}\curlywedge(\hat{Q}\curlyvee\hat{R})\natural[(\hat{P}\curlywedge\hat{Q})\curlyvee(\hat{P}\curlywedge\hat{R})]$, $\hat{P}\curlyvee(\hat{Q}\curlywedge\hat{R})\natural[(\hat{P}\curlyvee\hat{Q})\curlywedge(\hat{P}\curlyvee\hat{R})]$.
  Proof. We first prove that $A\cap(B\cup C)=(A\cap B)\cup(A\cap C)$ by Venn diagram (Fig. 4.1.) Then by two of the above dualities we get $\hat{P}\curlyvee(\hat{Q}\curlywedge\hat{R})\natural[(\hat{P}\curlyvee\hat{Q})\curlywedge(\hat{P}\curlyvee\hat{R})]$ and by the duality of $\curlyvee$ and $\curlywedge$ we get $\hat{P}\curlywedge(\hat{Q}\curlyvee\hat{R})\natural[(\hat{P}\curlywedge\hat{Q})\curlyvee(\hat{P}\curlywedge\hat{R})]$.
  The *caveats* of course apply: *A*, *B*, and *C* must be intensional sets and the disjunctions must be complete.

Theorems 4.6, 4.7, and 4.9 all illustrate that the larger an intension the smaller the extension that it defines. This is an indication that there are no infinite intensional sets, nor an intensional null set, since a null set would require an infinite

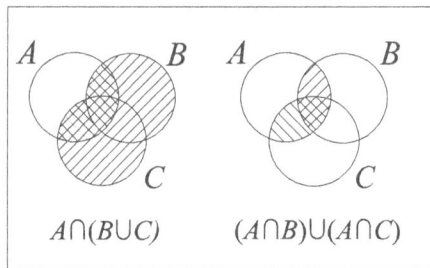

$A\cap(B\cup C)$ $(A\cap B)\cup(A\cap C)$

Fig. 4.1.

intension and an infinite extension would require a null intension (94).

Theorem 4.6, $(\hat{P} \succ \hat{Q}) \Leftrightarrow (\{\mathcal{E}\hat{P}\} \subset \{\mathcal{E}\hat{Q}\})$, shows the basis of extensional analytic truth (60) and of the idea that set-theoretic precursor truth is set membership (61).

Theorem 4.9, $\{\mathcal{E}(\hat{P} \wedge \hat{Q})\} \supseteq \{\mathcal{E}\hat{P}\} \cup \{\mathcal{E}\hat{Q}\}$, shows that a subset of an intensional set is not necessarily an intensional set: it may be only a contingent set.

Theorem 4.10 shows that if the disjunction is complete then $\wr \hat{s} \Rightarrow \wr (\hat{s} \wedge \hat{P})$: this is a triviality, and shows that the argument form called Disjunctive Addition, although valid, is harmless in relational set theory. The importance of this is that it limits, in relational logic, the extravagant irrelevance that this argument form allows in orthodox set theory and truth-functional logic: in $A \subset (A \cup Q)$ and in p→(p∨q), Q is any set whatever and q is any statement whatever, and so both are contingent and irrelevant even though their introduction is perfectly valid.

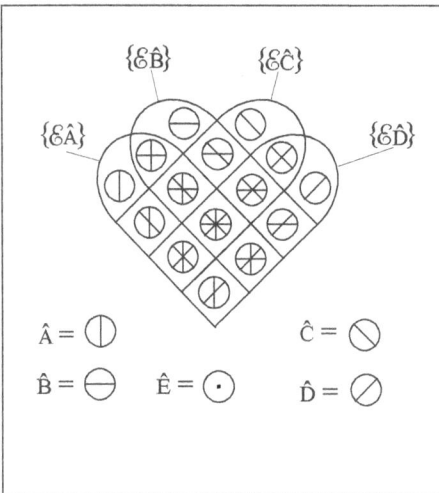

Fig. 4.2.

Relational meanings may be represented graphically, by **pattern diagrams**, in a manner complementary to the graphical representation of extensional meanings by Venn diagrams, also called Euler diagrams because (like so many other things) they were originally invented by Euler (1707-1783) but named after someone else. The use of patterns is natural enough: since patterns are relations, parts of the patterns represent properties of the relations. This is shown in Fig. 4.2, in which four intensions, Â, B̂, Ĉ, and D̂, are represented by four patterns; their intensional sets together form a quadruple Venn diagram, each set being shaped like an eight-paned roman-arched window. The commonality of the four intensions, $\wedge\{Â,B̂,Ĉ,D̂\}$, is represented by Ê, a circle (with a dot to represent the minim), and we assume that $\{ƐÊ\}$ is identical with the total quadruple Venn. Four of the theorems relating the relational and extensional connectives are illustrated in the diagram: conjunction, Theorem 4.7, $\{Ɛ(Â\curlyvee B̂)\} = \{ƐÂ\} \cap \{ƐB̂\}$; both kinds of disjunction, Theorem 4.9, $\{Ɛ(Â\curlywedge B̂)\} \supset$ $\{ƐÂ\}\cup\{ƐB̂\}$ — incomplete because $(Â\curlywedge B̂)ⅎÊ$ and $\{ƐÂ\}\cup\{ƐB̂\}\subset\{ƐÊ\}$ — and complete because $\{Ɛ(Â\curlyvee B̂\curlyvee Ĉ\curlyvee D̂)\} = \{ƐÂ\}\cap\{ƐB̂\}\cap\{ƐĈ\}\cap\{ƐD̂\}$; Theorem 4.6, implication, $(Â\succ Ê)$ and $(\{ƐÂ\}\subset\{ƐÊ\})$; and Theorem 4.5, equivalence, $(Â\curlywedge Ĉ)ⅎ(B̂\curlywedge D̂)$ and $\{Ɛ(Â\curlywedge Ĉ)\} = \{Ɛ(B̂\curlywedge D̂)\}$, because $(Â\curlywedge Ĉ)ⅎ(B̂\curlywedge D̂)ⅎÊ$ and $\{Ɛ(Â\curlywedge Ĉ\} = \{Ɛ(B̂\curlywedge D̂\} = \{ƐÊ\}$.

As another example of a pattern diagram, Fig. 4.3 illustrates incomplete disjunction. In this $(P̂\curlywedge Q̂)ⅎR̂$, as shown by their patterns, and $(\{ƐP̂\}\cup\{ƐQ̂\})\subset\{ƐR̂\}$, as shown by the Venn diagram, therefore $(\{ƐP̂\}\cup\{ƐQ̂\})\subset\{Ɛ(P̂\curlywedge Q̂)\}$.

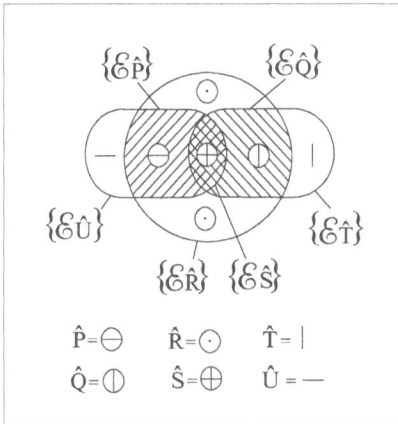

Fig. 4.3.

The Venn diagram also shows that P̂ and Q̂ are both false, meaning that $\{ƐP̂\}$ and $\{ƐQ̂\}$ have no members, as is shown by the shading; but the disjunction of P̂ and Q̂, as shown by their patterns, is their commonality, R̂: i.e. $(P̂\curlywedge Q̂)ⅎR̂$, and $\{ƐR̂\}$ does have members, so $(P̂\curlywedge Q̂)$ is true. In other words, with an incomplete disjunction the extensional disjunction can be false while the relational disjunction is true. We see this with the example of triangles and quadrangles (76): something may be neither a triangle nor a quadrilateral yet still be a polygon.

Notice, however, that the choice of patterns in pattern diagrams must be made carefully: in Venn diagrams a large choice of areas may represent a particular set, but in pattern diagrams the choice of patterns is more limited. A bad choice leads to nonsense.

We turn now to **purely extensional set theory**, or **contingent set theory**, which is relational set theory, less all intensions and all the purely nominal sets. It is, of course, ideal only, but not irrelevant in the present context.

In this theory a set can be defined only by enumerating its members. So any set too large to enumerate does not exist. It is usually assumed that $\{1, 2, 3, ...\}$ is an infinite enumeration of the set of all natural numbers, but it is not. The ellipsis, '...' is read as 'and so on' and contains a hidden intension, 'next successor', which defines every member except the enumerated first. So in general any use of 'and so on' is an implicit claim to the existence of an intension. So sets such P and Q cannot exist in this theory if they are too large, and *a fortiori* nor can their Cartesian product $(P{\times}Q)=\{\langle x,y\rangle: (x{\in}P)\wedge(y{\in}Q)\}$. Since such large sets are needed in defining extensional numbers (5, 98), this logic is unsuitable for that purpose.

A specific example of a large set will make this clear. Consider the set $\{\langle m,w\rangle: \langle m,w\rangle \subset \{\langle x,y\rangle \ni (x{\in}M)\wedge(y{\in}W)\}$, where M is the set of all men and W is the set of all women, and m is a particular man and w is a particular woman. (We are assuming, for the sake of this argument, that M and W do not have intensions but can be enumerated.) We can distinguish three sets that are subsets of this Cartesian product: $F=\{\langle m_1,w_1\rangle, \langle m_2,w_2\rangle, ...\}$, $H=\{\langle m_1,w_1\rangle, \langle m_2,w_2\rangle, ...\}$, $B=\{\langle m_1,w_1\rangle, \langle m_2,w_2\rangle, ...\}$, where the subscripts 1, 2, etc. do not necessarily indicate the same members in each of F, B, and H. F is the extension of the set of fathers of adult daughters, B

is the extension of adult brothers of adult sisters, and *H* is the extension of husbands of wives. Now consider a particular man and a particular woman, Bob and Alice, say: how can you decide whether, as a pair, they belong to *F*, *H*, *B*, or none of these three sets, using enumerations only? Let the ordered pair consisting of Bob and Alice be $\langle B, A \rangle$. You now have to examine every member of *F* to find out whether $\langle B, A \rangle$ is among them: if it is then Bob is the father of Alice; otherwise repeat for every member of *H*: if it is then Bob is the husband of Alice; otherwise repeat for every member of *B*: if it is then Bob is the brother of Alice. But the sets of men and of women are so large — billions of members, and larger still if they include dead or fictitious characters — that these searches are impossible. And there are a huge number of other possible relations between Bob and Alice: Bob might be a policeman giving Alice a ticket, a bank clerk cashing her check, a barman serving her a drink, a plumber fixing her kitchen sink, a priest hearing her confession, a pilot flying her to Athens, and so on. The classes can be distinguished easily by their intensions: fatherhood of a man and a woman, husbandhood of a man and a woman, or siblinghood of a man and a woman, so why should anyone bother with the elaborate definitions of relations as sets of ordered sets? The answer is that these elaborate definitions are attempts to define relations entirely by means of their terms — to make them "logical constructs" out of their terms — in order to make relations concrete rather than abstract; and this is impossible for relations that have emergent properties not possessed by their terms: the properties of fatherhood, husbandhood, and siblinghood cannot be found, or derived, exclusively from the properties of man and of woman: they *emerge* from suitable *configurations* of their terms, which is quite different from being constructed out of their terms. Such constructive attempts originate with nominalists, who deny the existence of anything abstract, hence of intensions: a denial that is doomed to failure since abstract entities — actual relations — are a fact of life.

Idempotence is our next consideration. In orthodox set theory expressions such as $A=A$, $A\cap A=A$, $A\cup A=A$, and $A\subseteq A$ are called idempotent and are allowed. That is, every set is identical with itself and the intersection, union, or subset of a set with itself is identical with itself. But here they cannot be — simply because there are no such monadic relations other than trivially or nominally: so '=', '∩','∪' and '⊆' cannot be monadic. This means that idempotence requires a plurality of terms, which is provided by similarity but not by identity: remember that similarity requires plurality while identity requires singularity (69): identity and similarity are disjoint. The equivalence theorem, $(\hat{A}ℛ\hat{B})\Leftrightarrow[(\{\mathcal{E}\hat{A}\}=\{\mathcal{E}\hat{B}\}]$, shows that idempotence may have relational meaning when applied to intensions, but not when applied to intensional sets. This theorem requires that there cannot be two similar intensional sets, for if $\{\mathcal{E}\hat{A}\}ℛ\{\mathcal{E}\hat{B}\}$ then $\hat{A}ℛ\hat{B}$, in which case $\{\mathcal{E}\hat{A}\}=\{\mathcal{E}\hat{B}\}$, by the equivalence theorem; this means that $\{\mathcal{E}\hat{A}\}$ and $\{\mathcal{E}\hat{B}\}$ are identical, they are one, hence not similar. So to say that idempotence, such as $\{\mathcal{E}\hat{A}\}\cup\{\mathcal{E}\hat{A}\}=\{\mathcal{E}\hat{A}\}=\{\mathcal{E}\hat{A}\}\cap\{\mathcal{E}\hat{A}\}$, is true requires that union and intersection may be monadic — and there are no such monadic relations; but $\hat{A}ℛ\hat{A}$ is meaningful because one instance of $\hat{A}$ is similar to another instance of $\hat{A}$. So relations such as =, ∩, ∪, and ⊆ have to be dyadic, and this requires that in the case of idempotence there must be more than one instance of their terms[8]. This happens with kinds of relations, such as $\hat{P}$, provided that there is more than one relation $\hat{P}R$; in such cases $\hat{P}ℛ\hat{P}$, $\hat{P}\curlywedge\hat{P}$, and $\hat{P}\curlyvee\hat{P}$ have relational meaning. This is significant mathematically in the case of equivalence relations, which are defined as any relation which is symmetric, transitive, and reflexive, and which play

---

[8] It might be thought that '=' in this context means set identity, and identity implies singularity, so the '=' in $A=B$ is monadic. But this usage is misleading: it does not mean equality, it means that the names A and B name the one identical set — and this one set cannot be either identical with, nor equal to, itself because there are no such relations.

an important role in mathematics. The property of reflexiveness requires idempotence, so equivalence relations are possible in relational set theory but not in exclusively extensional set theory. So in mathematics based on such extensional set theory equivalence relations are not possible, unlike mathematics based on relational set theory.

Arguing from an exclusively extensional point of view gives another way of putting this: there are no one-membered ordered sets, since an ordered set requires at least two members[9]; so in extensional set theory there are no monadic extensional relations, since relations there are defined by means of ordered sets; hence there are no reflexive relations, hence no equivalence relations, hence no idempotence. This limitation also applies to relational set theory in rare cases: those cases where only one instance is possible. For example, there is only one universe set, $U_{\hat{p}}$, of properties, so an expression such as $\hat{s} \, \wr U_{\hat{p}}$ has only nominal meaning. In Chapter 8 we will find another singular instance of a property.

It is important to distinguish between extensional set theory and purely extensional set theory. The former includes intensional sets and the latter does not.

Note that relational set theory has relational meaning, which has extensional meaning in the form of extensional set theory; and both set theories have nominal meaning in that they may be expressed nominally, in the form of symbols. Thus $R \Rightarrow E \Rightarrow N$.

---

[9] One might be tempted to define a special case of an ordered set, namely, a one-membered order. But since an order is a relation this special order would be a monadic relation, and there are none such because of their extravagant multiplication: every relation would be a one-order, including every one-order. We could allow the existence of relations called 'one-order relations' without them multiplying extravagantly, simply by denying them any uppers; but this would not explain anything, so is denied by Occam's Razor.

In **nominal set theory** there are names of sets that do not exist. One such is the empty set, symbolised by ϕ. It is usually defined as the set that has no members, and is included in set theory for maximum generality and maximum duality (244): it is a subset of every set, it is the intersection of disjoint sets, and it is the dual of U, the universe of discourse. But it may be equally well defined as the intersection of sets that do not intersect, or as a non-intersecting intersection, and so is self-contradictory, hence impossible, hence non-existent mathematically. Other such purely nominal sets are all the infinite sets. Justification for this claim is given in Chapter 5 (94), but here may be based on the remark that the definition of infinity does seem to produce many paradoxes; and both paradoxes and contradictions have only nominal meaning. Also, to define something, such as "Infinity is the number of numbers", does not automatically give it existence.

Chapter 5. Relational Mathematics.

Having examined the relational, extensional, and nominal foundations of logic and set theory, we now do the same for the foundations of elementary arithmetic and geometry. We are concerned primarily with applied mathematics, since we are concerned mainly with the relations between mathematics and science, and between these and the empirical and noumenal worlds.

We begin with some general points. Axiom generosity — the cornucopia of theorems coming out of an axiom set — is emergence of theorems; only mathematics has it; and only relations emerge. Thus theorem-rich mathematics is relational. Mathematics includes the relation of equality and hence the concept of equations, as well as other equivalence relations such as congruence and isomorphism. Mathematics includes genuine functions — necessitous functions — which are relations. And mathematics includes the concept of a variable, which can only be defined by means of an intension. Some of all this may be defined within set theory, but only by intensions of sets, which require relations. So mathematics is essentially relational, not set-theoretical.

We next need two more definitions: first, if a relation R has a property set $\hat{r}$ then R generally is a **representative instance** of a property set $\hat{s}$ if $\hat{r} \divideontimes \hat{s}$ (73), and specifically so if $\hat{r} \succ \hat{s}$. This is, of course an ancient usage: Euclid, needing to prove something about triangles in general, would let a particular, drawn, triangle be a representative instance of triangularity, because triangularity *per se* is not a relation — it is a property of a relation — and cannot be drawn; and of course no property of the particular triangle other than triangularity would be considered in the proof. This concept is also needed when $\hat{s}$ is a set of properties that does not contain the minim, $\acute{m}$, so that $\hat{s}$ does not define a relation; this happens when needing to discuss one or more properties independently of any relation that might possess them.

Secondly, we define the **subordinate levels** of a relation: if the emergent level of a relation R is n, $\#R = n$, then the emergent level of the terms of R, n-1, is the first subordinate level of R, the emergent level of the terms of the terms of R, n-2, is the second subordinate level of R, and so on.

We have seen that every relation, without exception, has a term set, and the three properties of possibility, simplicity, and an adicity: relations without these are impossible, merely nominal. And we also noted that the adicity of a relation is the number of terms that it has, the number of members in its term set. We now state this formally:

Except for the number one, a **relational natural number** — that is, a natural number that has relational meaning (10, 40) — is an adicity. As such, a relational natural number is a single property of a relation, the adicity of that relation.

The **relational number one** also comes from the minim: but from simplicity, not from adicity. The simplicity in the minim means that every relation is indivisible, it has no parts; so simplicity is essentially unity. Every relation possesses the minim and so is a representative instance both of an adicity and of the number one, as is shown by the number of its term set and it being one relation.

More precisely, an n-adic relation, R, is a representative instance of the number n, the number n itself being the intension of the set of all adicities that are equiadic (see next page) with R — and this intension is $\aleph n$; but since $\{\mathcal{E}\aleph n\}$ is a similarity set (49, 75), every member of $\{\mathcal{E}\aleph n\}$ is an intension of $\{\mathcal{E}\aleph n\}$ and so (93) the adicity of R is the number n as well as being a representative instance of n.

In the following 'natural number' and 'number' will refer to 'relational natural number' unless otherwise stated.

We will symbolise the natural number of a relation R by the same letter, lower case, italic: r. That is, r is the adicity of R; this is symbolised ^rR. Because r is a property we strictly should symbolise it as a subset of $\hat{r}$, given $\hat{r}R$; but we will generally stay with the more conventional usage, r, for numbers, and use $\hat{R}$ as the adicity of R when needed.

Every relation is a representative instance of its own adicity and of its simplicity.

Two relations are **equiadic** if their adicities are similar.

Two relational natural numbers m and n are **equal**, symbolised $m = n$, if they have representative instances M and N which are equiadic.

Obviously, the number of the term set, R, of a relation R is the adicity, r, of R; thus we have the meaning of the relational number of a term set, and hence the number of a set: the **relational number of a set** S is the number of any term set that is in one-to-one correspondence with S.

The relation **greater than**, along with its inverse **less than**, which holds between any two numbers that are not equiadic, are next defined. Using the proper notation, the numbers n and m are $\hat{N}$ and $\hat{M}$ and n is greater than m, symbolised $n>m$, or $m<n$, if and only if $\hat{N} \succ \hat{M}$: $(n>m) \Leftrightarrow (\hat{N} \succ \hat{M})$.

Addition of two numbers, n and m, symbolised by $n+m$, is their coupling: $(n+m)=(\hat{N} \curlyvee \hat{M})$. This is called the **sum** of n and m.

We can now define specific numbers by means of addition of the number one, and order them by means of the relation of greater than. Thus we begin with the number *1* and add *1* to it, and repeat:

$$1+1 =_{\text{def}} 2$$
$$1+1+1 =_{\text{def}} 3$$
$$1+1+1+1 =_{\text{def}} 4$$
$$1+1+1+1+1 =_{\text{def}} 5$$

and so on; and *2<3<4<5...*

We can next show, using parentheses for clarity, that:

$$5 = (1+1)+(1+1+1) = 2+3$$
$$5 = (1+1+1)+(1+1) = 3+2$$
$$5 = 1+(1+1+1+1) = 1+4$$
$$5 = (1+1+1+1)+1 = 4+1$$

so that *2+3=3+2* and *1+4=4+1*. This can be generalised to
n+m=m+n, using mathematical induction, and the
commutation of arithmetical addition thus proved. Association
of addition also is proved similarly.

Products and powers of numbers now follow in the
usual fashion. The **multiplication** of two numbers, *n×m* is
repeated addition and the **power** of *n* to the *m*, n^m, is repeated
multiplication:

$$n×m = n_1+n_2+n_3+...n_m.$$
$$n^m = n_1×n_2×n_3×...n_m.$$

The inverses of these operations are subtraction,
division, and the taking of roots. Subtraction of *m* from *n* is
defined as N̂-M̂, the decoupling (74) of *m* from *n*. Division is
then repeated subtraction and the taking of roots is repeated
division. Note that since N̂-M̂ excludes the possibilities of N̂=M̂
and N̂<M̂ the relational natural numbers cannot include zero or
negative numbers.

An equivalent definition of addition of relational
numbers is the compounded relation of join (42) between all
the terms of a relation. Thus with an *n*-adic relation, ^nR, each
of its terms is a relation, so a unity, and so is a representative
instance of the number one; with joins between these terms we
get $1_1+1_2+...1_n = n$. Thus a dyadic relation has *1+1* terms, a
triadic relation has *1+1+1* terms, and so on. Such a set of
additions is a special case of a sum; it is part of the
configuration (39) of the terms of R. A more general case of a
sum is the join of the adicities of the terms of a relation, rather
than the join of the terms as representative instances of the
number one. If we define the **first subordinate adicity** of a

relation R as the sum of the adicities of the terms of R then if R has the *m* terms $^aR_1$, $^bR_2$, ... ^nR_m, where the superscripts are their adicities, then the first subordinate level adicity of R is *a+b+...n*. Similarly, the second subordinate level adicity of R is the sum of the adicities of the terms of the terms of R, and so on for lower subordinate levels. A special case of the subordinate adicity of R is where all the terms of R are equiadic. Thus if R is *m*-adic and each term of R is *n*-adic then the first level subordinate adicity of R is $n_1+n_2+...n_m$. This is the product of *n* and *m*, or the **multiplication** of *n* by *m*. The commutation of multiplication can easily be proved, and we can define an operation or function called multiplication.

These definitions in relational arithmetic lead to seven observations concerning relational natural numbers.

1. The similarity set (49, 75) of a natural number is the set of every relation that is equiadic to it. It follows that there is only a nominal difference between a particular number and various instances of it: the number is the intension of the similarity set and the unified totality of instances of it is the similarity set, but each member, with ℵ, serves as an intension of the set. Thus in an expression such as $2+2 = 2 \times 2 = 2^2$, each of the six instances of *2* is the number 2, since, with ℵ, it serves as a dyadic intension of the similarity set of all dyadic relations. Note that it is the intension that is the number two; the similarity set itself is the extensional meaning of the relational number two, as opposed to the extensional natural number two, which is the set of all extensional sets that are in one-to-one correspondence with the term set of a dyadic relation.

2. There is no relational natural number zero: the word *zero*, as a number, has only nominal meaning. We could define it as the adicity of any "nonadic" relation, but none such exist — they are at best exclusively nominal relations. Or we could

define it as the extensional number (5, 98) of the null set, but there are no null sets in intensional or exclusively extensional set theory: the null set, which is the intersection of non-intersecting sets or the extension that is not an extension, is self-contradictory and so has only nominal meaning. However, zero does have some relational meaning: it is a place holder in our decimal notation; a place holder between $+1$ and -1 on the number line; and a placeholder for the value of $n-n$. And a placeholder, even though it is not a number, is a relation. The fact that zero is not a number is shown by it threatening nonsense when assumed to be a number — as with division by zero and the definition of 0^0.

3.   There are no relational infinities. The standard definition of an infinite number is that it is the number $\aleph$ (aleph) of a set which possesses at least one proper subset having an equal number, $\aleph$, of members. If M and N are two such intensional sets, such that $N \subset M$ (that is, $N \neq M$; N is a proper subset of M), having members n and m, such that $n = m$, then this latter requires that $N=M$ hence $N \not\subset M$. This contradiction requires that $\aleph$ has only nominal meaning[10].

Also, Zeno's paradoxes, which were designed to show the impossibility of motion, given infinite divisibility, in fact show by *reductio* that because motion does in fact exist, infinite divisibility is impossible.

Another argument against relational infinity is that of the intuitionist L. E. J. Brouwer (1881-1966). The number of natural numbers was symbolised by Cantor (1845-1918) as $\aleph_0$, and it is true of $\aleph_0$ that $(\aleph_0 - n) = \aleph_0$, where n is any finite natural

---

[10] Bertrand Russell, in discussing the one-to-one correspondence between the set of natural numbers and the set of even numbers, wrote: "Leibniz, who noticed this, thought it a contradiction ... Georg Cantor, on the contrary, boldly denied that it is a contradiction. He was right; it is only an oddity." (*A History of Western Philosophy*, Allen and Unwin, London, 1946, p.858.) As much as I admire Cantor and Russell, I have to agree with Leibniz.

number. It follows that if n is a finite natural number then $n+1$ is a finite natural number, because if $n+1$ is infinite, $(n+1) = \aleph_0$ and then $n = (\aleph_0 - 1) = \aleph_0$. Equally, if n is finite then so is $n+m$, where m is any finite natural number. And since multiplication is repeated addition, $n \times m$ is finite. So is n^m. Thus every finite number of arithmetical operations on every finite number of finite natural numbers yields a finite natural number. So $\aleph_0$ is not a natural number. So we may say that, relationally, all arithmetical operations on $\aleph_0$, or with $\aleph_0$ itself on finite numbers, are not defined; equally so with extensional numbers because there are no infinite enumerations, so all have nominal meaning only. Or, to put this another way, $\aleph_0$ has no operational definition; or, as Brouwer put it, infinity cannot be constructed. Consequently $\aleph_0$ has nominal meaning only.

Again, a finite mind cannot contain an infinite-adic relation, so such a relation has no ideal existence. It also has no empirical existence (135): we cannot perceive it. And the only ground for claiming noumenal existence (28) for it would be that it has value in explaining something in the empirical world, and it explains nothing empirical. So an infinite-adic relation, and hence $\aleph_0$, have nominal meaning only.

The denial of infinity requires that there be a greatest natural number, as explained in the next paragraph.

4.   If there are no relational infinities then there must be a greatest finite relational natural number: the number of terms, and all-level subordinate terms, of the top relation in the largest whole, as will become clear in Chapter 8. Call this number g. Then if $m < g$ and $n < g$ are relational natural numbers then $n+m$ is a relational natural number only if $(n+m) \leq g$; and if $(n+m) > g$ then $n+m$ is an exclusively nominal number — because not only is it not a relational number, it is also not an extensional number because it cannot be enumerated if g cannot be enumerated. Thus there is no closure on relational addition or multiplication. If the relational number g is large

beyond our comprehension, this non-closure is of no practical significance.

5.   The essential difference between pure and applied mathematics is that pure mathematics is concerned with every *possible* relation and applied mathematics is concerned with every *actual* relation. See Chapter 8 for more on this difference. Since the truth of relational arithmetic is relative to reality (as well as being relationally analytic), relational mathematics is exclusively applied mathematics. This is not to denigrate pure mathematics, which, although containing some exclusively nominal meaning, is undoubtedly mostly relational — as shown by its axiom generosity, necessities, and beauty — but this relational meaning either is a so far unidentified part of applied mathematics or else is relational meaning independently of the relational meaning of applied mathematics. Another way of putting this is that pure mathematics studies possible noumenal worlds and applied mathematics studies the actual noumenal world.

6.   We may note that Godel's (1906-1978) theorems, on the incompleteness and consistency of any system large enough to contain number theory, are not proved in a purely relational arithmetic, for three reasons: first, his proofs apply to formal systems only such as Whitehead (1861-1947) and Russell's (1872-1970) *Principia Mathematica*, and relational mathematics is not such; second, because his proofs are essentially extensional and rely on self-reference which has no extensional meaning — since idempotence (86) is not extensionall possible; and third, because relational arithmetic is finite, so it cannot necessarily be large enough to map the requisite formulas into itself. In fact, since inconsistent mathematics has nominal meaning only, it follows that purely relational arithmetic must be consistent. Also, as we shall see in Chapter 8, relational applied mathematics must be complete as well as consistent.

7.   On relational natural numbers, we may paraphrase Leopold Kronecker's (1823-1891) famous dictum and say that God made the relational numbers and that all the extensional numbers, all the nominal numbers and all the other numbers — negative, rational, real, imaginary, complex, etc. — are the work of man. One could define relational rational numbers as ratios, but from a relational point of view a ratio is not a number — that is, not an adicity. One could also define negative numbers by means of the sense of vectors (101), and imaginary numbers by defining *i* as a versa, the anticlockwise rotation of a vector through a right angle, but relationally these are not adicities. The key point here is that to define something mathematically does not ensure that it has relational meaning — although it does ensure nominal meaning. This is not to say that negative, rational, and imaginary numbers have no relational meaning: they do have relational meaning, but they are not relational numbers because they are not adicities. Also, to define principles of closure, on addition, multiplication, subtraction, division, the taking of roots, etc., does not ensure that these principles are relationally true.

One further point concerning relational mathematics: in contemporary mathematics: the concept of function includes two kinds of function, which may be defined as **relational functions** (110) and **contingent functions**. Each has a domain and a range and is a function because it relates each member of its domain to only one member of its range. A relational function does this necessarily, because it is a necessity, which is *singular* possibility (7, 42, 43), which guarantees that it relates each member of its domain to only *one* member of its range; while a contingent function has to have these values enumerated. Thus the trigonometric functions are relational functions because $\tan(\pi/4)$, for example, is necessarily the number one, and this can be proved; but the function relating a guest list to their seating arrangements is a contingent function because where each guest sits is contingent and so has to be enumerated. Functions as defined in modern mathematics are **extensional functions**: they are either relational functions or contingent functions. They should be relational functions only, since contingent functions serve no purpose in mathematics other than to give an illusion of maximum generality.

We turn next to extensional and nominal arithmetic.

We begin by distinguishing the extensional meaning of a relational number from the relational meaning of an extensional number.

A relational number n is the adicity of an n-adic relation, as we have seen.

The **extensional meaning of a relational number** n is the term set of an n-adic relation, which is an n-membered intensional set. The extensional meaning of the relational number one cannot be defined in this way, so is defined as the extension of the set of all instances of the relational number one. Conversely, the **relational number of any set S** is the number of any term set that is in one-to-one correspondence (5) with S. The relational number of a contingent set is obviously a special case of this.

An **extensional number** n is the set of all n-membered *extensional sets*; that is, the set of all sets, *intensional* or *contingent*, that are in *one-to-one correspondence* with an n-membered extensional set.

The **relational meaning of an extensional number** n is the intension of that number; that is, the intension of the set of all n-membered extensional sets.

Extensional arithmetic now follows in the usual set-theoretic way. For example, if two disjoint sets M and N have extensional numbers m and n then the sum, $n+m$, of n and m is the extensional number of $N \cup M$, which is the set of all sets that are in one-to-one correspondence with $N \cup M$.

The definitions of relational arithmetic are both more simple and more beautiful than those of extensional arithmetic. The significance of simplicity and beauty, in mathematics, science, and philosophy, is discussed in Chapter 7.

A **nominal number** is the name of any relational or extensional number for which a name or symbol exists, or any other number defined nominally.

Nominal arithmetic is the arithmetic of nominal numbers. It includes numbers that have exclusively nominal meaning. One such is the number zero which, as already explained (93), as a number, has exclusively nominal meaning. Another class of numbers that have exclusively nominal meaning is the transfinite numbers. They do not have relational meaning, as already explained (94), so unless it can be shown that there are contingent sets which have an infinite number of members, there is no alternative but to say that infinities have exclusively nominal meaning. But contingent sets can only be defined by enumeration, and an infinite enumeration is impossible. So there are no infinite contingent sets. Hence all infinite numbers are exclusively nominal.

We next consider relational geometry, which begins with the concept of axiom relation, or separator.

As we saw in Chapter 1, relations at the lowest level are separators called **axiom relations** (24). Axiom relations are compoundable relations (38); they have properties called **unit magnitude**, or unit measure, which sum. Compounded relations formed from them have magnitudes of the same kind, equal to their adicity. The alternative to having axiom relations is an infinite regress (140) of lower and lower levels. One reason for denying such an infinite regress of levels is that axiom relations need to have compoundable properties which sum: properties such as length and angle, out of which a geometry should emerge. If the axiom relations that possess the magnitude called length were geometric points, which is what an infinite regress would require, then they would not have a magnitude that sums — since a point has no magnitude. Another reason that points could not be separators is that no point in the traditional continuum has a point next to it because between any two such points therein there is an infinity of points, so that a point cannot immediately separate two other points; thus traditional geometric points cannot be relations or properties of relations, so cannot have relational meaning. Another reason for denying an infinite regress is that variety (132) of kinds of relation diminishes with level number and cannot be less than two; so there must be a lowest level.

We do not know what the axiom relations of the noumenal world actually are, other than being separators. The depictions of them here are merely suggestions, illustrations of possibilities.

Axiom relations provide, in principle, a foundation for both geometry and applied mathematics. We begin with geometry. We assume two kinds of axiom relation, each possessing magnitude:

A **linear separator** has unit length.

An **angle separator** is perhaps a right angle, perhaps a radian (because these are natural, like the Planck units (see Appendix D)), perhaps either, or perhaps simply elastic; or even very small, so that other angles (which might be Diophantine angles) are integral compounds of them. A right angle is assumed here for ease of exposition.

These separators may also be thought of as atomic lengths and angles — using *atomic* in its original Greek sense of an indivisible. If four linear separators and four right angle separators are combined so that each of the linear separators separates two of the angle separators, and each of the angle separators separates two of the linear separators, in one plane, then there emerges a new relation: a level-2 atomic square, having the emergent property of atomic area (131). Level-3 atomic cubes and volumes emerge similarly, out of six atomic squares and twelve angle separators. Atomic areas are **emergent separators** because they separate other atomic areas, as atomic volumes also are emergent separators separating other atomic volumes; but they are not atomic separators, like axiom separators, because they are not in the axiom level and they do not have to form separator loops (31). Atomic lengths, areas and volumes are also compoundable relations that sum, so that other lengths, areas, and volumes are sums of integral numbers of atomic lengths, atomic areas, and atomic volumes. Thus out of these separators we get a geometry on a nominal orthogonal lattice. Angles other than right angles might emerge with Diophantine right-angled triangles, in which the emergent hypotenuse and the other two sides all have lengths which are an integral multiple of an atomic length, and in which there are two emergent angles — between the hypotenuse and the other two sides. Clearly, as the length of a linear separator decreases towards the limit of zero, this macroscopic geometry approaches the limit of a Euclidean geometry on the continuum; but the limit itself has exclusively nominal meaning. This suggests that there are no relational circles, only approximations to them. At the physical atomic scale of 10^{-15}m the nearest approximation to a circle could be a regular polygon having 10^{20} sides of length l_p, the Planck length — since the order of magnitude of l_p is 10^{-35}m (see Appendix D). But note that if the atomic lengths were somewhat elastic, each of the 10^{20} of them could be strained to a smooth curvature, so as to convert the polygon to a circle.

Two further kinds of separator are needed to round off this picture. A **temporal separator** has unit duration, and is asymmetric. It is an atomic duration. An **atomic vector** has length, like a linear separator, but a length slightly less than that of a linear separator; so a structure of them forming an atomic volume has unit mass, through stressing space according to an inverse square law. This requires that atomic lengths and atomic angles are somewhat elastic — a requirement that does not deny their atomicity but does allow gravitational waves; if the atomic separators were a Planck length and a Planck time then the velocity of these waves would be c, the velocity of light. An atomic vector is fundamentally dissimilar to an atomic length because it has a sense and so is asymmetric.

Thus we have in principle the three basic units of dimensional analysis (see Appendix D): mass, length, and time.

A uniform expanse of atomic lengths does not contain any configurations leading to emergent relations. This makes clear that an emergence-configuration in the axiom-level can only be defined with dissimilarity relations. The emergent square, above, has eight dissimilarities in its configuration: one between the terms of each pair of length and angle. Because of this there are no purely arbitrary lengths, made up of an arbitrary number of atomic lengths: compound lengths always consist of a number of atomic lengths bounded at each end by dissimilarities, as do compound areas and volumes. This applies to all compoundable relations. As remarked earlier (33), a boundary is defined by dissimilarity relations.

Chapter 6. Three Kinds of Meaning.

This chapter reviews the three kinds of meaning —
relational, extensional, and nominal — that arise from making
relations primitive and defining numbers, sets, etc. in terms of
them. To repeat their definitions: relational meanings are
relations and properties of relations, and they include
inherence, emergence, and demergence, and intensions of sets.
Extensional meanings are sets and their members and
extensions — a set being a set relation (48) defined by an
intension and the relation-every, an extension being the term
set of a set relation, and a member, being defined by an
intension R_T, as any coterm of T in R. Extensional meanings
include contingent sets, which are sets that do not have an
intension but can be defined by enumeration. And nominal
meanings are words or symbols, as opposed to the denotations
of these, and they include the words or symbols for relational
and extensional meanings. Relational meaning allows axiom
generosity (40) and is consistent; nominal meaning allows
paradoxes and contradictions, and therefore may be
inconsistent; and extensional meaning allows neither
generosity nor inconsistency.

To avoid confusion we distinguish between, on the one
hand, relational, extensional, and nominal meanings; and on
the other hand, relational, exclusively extensional, and
exclusively nominal meanings. The former are such that each
necessitates the next: $R \Rightarrow E \Rightarrow N$. The latter are such that
exclusively nominal meaning excludes extensional and
relational meaning, and exclusively extensional meaning
excludes relational meaning. The word *purely* in this context
means *exclusively*.

There is also a possible confusion when relational
meaning exists without either extensional or nominal meaning:
a creative thinker may come up with an original abstract idea
— a relational meaning — for which no extensional or
nominal meaning yet exists. But this last does not need further
discussion, since a name or symbol may be easily assigned to

the idea, to make a concept; and extensions arise automatically from the idea because it is a relation and so defines natural sets, or else it is a property of a relation and so defines the property set of a kind of a relation.

We next look at eleven illustrations of the three kinds of meaning.

1.   Relations and their properties *are* relational meanings, but relations also have extensional meanings and nominal meanings. Exclusively extensional meanings of relations are extensions of intensional sets, or subsets of a Cartesian product (5, 84). Exclusively nominal meanings of relations are polyadic predicates or relational names or descriptions that have no reference. A **polyadic predicate** (237) is a description that occurs in modern logic (63, 231): some logicians suppose that grammatically a subject may have one predicate or many predicates. This gives rise to what are called the monadic predicate calculus and the polyadic predicate calculus, two developments of truth-functional logic that include the flaws of this logic.

2.   Truth — analytic, synthetic, and precursor — usually has the three kinds of meaning. Relational analyticity (54) is superintension, extensional analyticity is subset (60), and nominal analyticity (63) is tautology, which is demonstrated if its denial leads to a contradiction. Relational analytic truth (54) is based on necessity. Extensional analytic truth (60) differs from relational analytic truth because of the implication theorem (79): $(\hat{A} \succ \hat{B}) \Leftrightarrow (\{\mathcal{E}\hat{A}\} \subset \{\mathcal{E}\hat{B}\})$; since superintension is relational analyticity, subset is extensional analyticity (60). Nominal analytic truth (63) is consistency, and nominal analytic falsity is inconsistency, as shown by contradiction or paradox. Relational and extensional synthetic truth (52) is similarity of a concept, proposition, set-intension, or number of a set to either empirical reality or to noumenal reality, a concept being an abstract idea, which is an ideal relation or

property thereof, bonded to a word or symbol, and a proposition being a structure of concepts. This truth is sometimes called **correspondence truth**: to varying degrees (182) a relational idea, proposition, or theory may be similar to reality — empirical or noumenal (53) — and thereby true to that degree. Nominal synthetic truth is correct (i.e. established) usage of language, correct statement of meaning — relational, extensional, nominal, ideal, noumenal or empirical meaning.

There is also precursor truth. Relational precursor truth (55) is the property of possibility that is in the minim, and so a property of every relation; this property is equally consistency and logical existence. This precursor truth is the basis of existential presupposition (56). Extensional precursor truth (61) is set-membership: a set is true in this sense if it has members, and false otherwise. So a set-theoretic universe of discourse is an extensional precursor tautology and the null set is an extensional precursor contradiction. Nominal precursor truth (63) is sense, as opposed to the nonsense that is nominal precursor falsity.

Since we are cataloguing the forms of truth, two more may be mentioned here. **Coherence truth** is richness of necessities in an explanation or, equivalently, absence of arbitrariness; as such it is primarily relational. And there is also **empirical truth** and falsity: **empirical falsity** is illusion, and empirical truth is non-illusion: the bentness of the half-immersed stick in water is empirically false because the stick is really straight.

These ten meanings of the word truth all have their opposites, yielding ten meanings of the word false. Thus relational falsity may be absence of necessity, or else dissimilarity to reality, or else absence of possibility. Extensional falsity may be dissimilarity to reality, absence of subset, or else non-membership. And nominal falsity is incorrect usage, as in error or deceit. There are also three forms of falsity that yield truth when denied: these are the three meanings of complementarity: relational complementarity is dissimilarity, extensional complementarity is set complementarity, and nominal complementarity is the use of negative prefixes such as 'non', 'un', and 'dis'.

Given that the null set is the extensional meaning of contradiction, there is a set-theoretic version of the well known feature of truth-functional logic that from a contradiction you may derive q, anything you please. We saw the proof of this on page 65. The extensional analogue of this is:

1. $\wr(a\epsilon\phi)$ Premise.
2. $a\epsilon(\phi\cup Q)$ 1, Disjunctive addition.
3. $(\phi\cup Q)=Q$ 2, Definition of ϕ.
4. $a\epsilon Q$ 2, 3, Equivalence.

Note that to say that $a\epsilon\phi$ is to say that ϕ is true, because it has a member; and this is like saying that a contradiction, such as $p\wedge\sim p$, is true. This argument is a valid reason for denying the existence of the null set in set theory, other than to show error.

3.   A particularly important relational property is necessity, which occurs in logical implication, causation, inherence, emergence, demergence, and relational equations and functions; this **relational necessity** is singular possibility, and is the relational meaning of the word necessity. For example, relationally, given the truth of an equation and the truth of one side of the equation, the truth of the other side is a singular possibility.

Extensional necessity, the extensional meaning of the word necessity, is universality — as occurs with subsets: if the contingent set A is a subset of the contingent set B, then members of A are universally members of B, but only contingently so, hence not relationally necessarily so. Empiricists sometimes define causation as universal correlation — one event is the cause of another if they are *always* correlated, their correlation is universal; but, unfortunately for empiricists, this universality cannot be perceived so is non-empirical. (Universal correlation is of course empirical evidence for noumenal causation, which does have relational necessity, which is singular possibility.)

Nominal necessity is tautology and a tautology is a statement or an argument that is always true, even if it has exclusively nominal meaning, as in the truth-functional statement that from a false proposition you may validly infer q, anything you please: $\sim p \rightarrow q$. Nominal necessity also includes purely nominal necessities such as those of superstition, stereotypical beliefs, and other false generalisations: such necessities exist only in language.

The difference between relational and extensional necessity is of major epistemological significance in science: extensional necessity is limited by the **problem of induction**, relational necessity is not. This problem is the problem of the invalidity of arguing from some to all, as in "All swans are white" — once a famous example of reliable induction, proceeding from all swans so far seen being white to every swan being white, and then falsified by the discovery of black swans. Superstition and stereotyping both exemplify the weakness of this type of generalisation. The problem of how to justify induction is the major problem in the philosophy of empirical science: how does one justify inferring a universal law from all known instances? But relational necessity, which appears in mathematics and relational logic, does not need induction. (Mathematical induction, which is a necessitous inference, is not induction in this sense.)

In technical language, if it is known that x is true of every member of *S*, and $S \subset P$, then it cannot be inferred validly that x is true of every member of *P*. But if *S* and *P* are necessary sets, $S=\{\mathcal{E}\hat{s}\}$ and $P=\{\mathcal{E}\hat{p}\}$, then $(S \subset P) \Rightarrow (\hat{s} \succ \hat{p})$, by Theorem 4.6, and what is true of $\hat{s}$ is true of $\hat{p}$, analytically; while if S and P are contingent sets this does not apply. In empirical science *S* and *P* are contingent sets; in theoretical science $\hat{s}$ and $\hat{p}$ are necessary sets. Thus the problem of induction shows the major epistemological difference between relational and extensional necessity.

However theoretical science, which is relational, also has an epistemological difficulty: it is speculative, hence never certain. One could say that theoretical science is speculative regarding synthetic truth while empirical science is speculative regarding analytic truth. These problems are discussed further in Chapter 10 (165).

Note that singular possibility entails universality, which entails tautology: R⇒E⇒N.

4.   In logic the connectives are defined relationally by similarity, extensionally by identity, and nominally by truth-functions, as we saw in chapter 4. The most important connective is implication, which relationally is superintension (and sometimes inherence, emergence, or demergence), extensionally is subset, and nominally is truth-functional material implication (63, 232) — which latter is false only if the antecedent is true and the consequent false. Treating them all in this order, equivalence (57, 58) is symmetric implication which is similarity of intensions, identity of extensions (71), and similarity of truth values (232). Conjunction is coupling of intensions (73), intersection of sets (74), and nominally true only if both conjuncts are true (231). Disjunction is commonality of intensions (73), union of sets (73), and nominally false only if both disjuncts are false (231). R⇒E⇒N applies to each of these triads of definitions.

However, more needs to be said on conjunction and disjunction. With the exception of the relational cases, they may all be defined truth-functionally: a conjunction is true if both conjuncts are true, and is otherwise false, and a disjunction is false if both disjuncts are false, and is otherwise true. The reason for this comes from the extensional cases of conjunction and disjunction. In order to explain this we will use the symbol ϕ for extensional falsity, which is extensional precursor falsity, or non-membership. This symbol is the standard symbol for the null set, which has no members and has nominal meaning only. But we do need such purely nominal symbols or words, in order to discuss and correct falsity. And the significance of ϕ here is its use with extensional conjunction and disjunction: for any membered set S, we have $(S \cap \phi) = \phi$ and $(S \cup \phi) = S$. So, since ϕ is always false, extensionally, and S is true because it has members, an extensional conjunction, or intersection, is false if at least one of its conjuncts is false, and a disjunction, or union, is true if at least one of its disjuncts is true. Arguably, because of the human propensity to think with classes — we are born to classify — this is the origin of the everyday grammatical use of conjunction and disjunction; and it is the major rationalisation for believing that Boolean algebra represents logical thought (67). All of this applies equally to compound conjunctions and disjunctions.

On the other hand, relational conjunction and disjunction are not definable truth-functionally, which should not be surprising: truth and falsity are ideal only, they apply to thought, imagination, and belief, and hence to language — while the noumenal is relational only, as shown by the Grand Structure (8). So relational conjunction and disjunction remain simply commonality and coupling.

More loosely, the connectives are relations. In relational logic they are relations between kinds, in extensional logic they are relations between instances, and in nominal logic they are relations between truth values.

In truth-functional quantificational logic $(x)\,\Phi x$ signifies universality, which is extensional necessity, and $(\exists x)Px$ means that P has at least one member, which is extensional existence, or extensional possibility; thus truth-functional quantification is based on extensional meaning, not on relational meaning.

5.　　In set theory, relational set theory (69) is the theory of *intensions* and their connectives, relationally defined *extensions* and their connectives, and of the *relations* between these two kinds of connectives; extensional set theory (84) has this plus contingent sets, and nominal set theory (88) has all this plus null sets and infinite sets. Relational set theory has axiom generosity, nominal set theory has contradictions, and extensional theory has neither of these. If we deal alternatively with exclusively extensional set theory (87) and exclusively nominal set theory, then extensional theory has no intensions, necessary sets, null sets, or infinite sets but it does have contingent sets, while exclusively nominal set theory has only purely nominal sets, such as the null set and infinite sets, and so is not a theory.

6.　　　In mathematics relational numbers (90) are adicities. Extensional numbers (5, 98) are sets of sets, where the latter sets are all in one-to-one correspondence; the set of sets that is an extensional number may only be defined intensionally since it is too large to be enumerated, and its intension is what all the sets in this set have in common. Nominal numbers (99) include relational and extensional numbers and also include infinity and zero, which the relational and extensional numbers do not. Infinity and zero both invite paradox, which is indicative of purely nominal meaning.

7.　　　A **relational function** is any ideal relation whose property set is a superintension of the property of necessity. Each of its possible antecedents is called an argument of it, and its singular consequent, for each argument, is called the value, or image, of that argument. These are called antecedent and consequent in the case of implication, and cause and effect in causation. The set of every possible argument of a function is called its domain and the set of every possible value of it is called its range. A relational function is relational because each argument necessitates its value. A relational function may be specified by a rule, since a rule specifies the intensions of the domain and range, and the necessity between each argument and value.

Like all relations, instances of functions are determined by their terms — argument and value — and kinds of functions are determined by their property sets. But the kind of relation called a relational function is determined by the commonality of the property sets of all relational functions, which is the property of necessity.

It was earlier (7) pointed out that we only know of causations by analogy with logical necessities. These necessities are those of relational functions, and if the functions themselves are applicable to the noumenal world then they describe these causations, as explained in Chapter 7.

A **contingent function** between two contingent sets, S and T, is a set of assignments to each member in S of a single value in T. S is called the domain of the function set and T is called the range of the function. We note that an assignment is a relation, usually created by a mind. Two examples of contingent functions are **extensional relation-every** and **extensional relation-any**, which are the extensional meanings of the relation-every, $\mathcal{E}$, and the relation-any, $\mathcal{A}$. Extensional relation-every is the universal quantifier of truth-functional quantificational logic, and extensional relation-any is the existential quantifier of truth-functional quantificational logic, or else a random selection of a member of a contingent set.

An **extensional function** is a set of ordered pairs determined by either a relational function or a contingent function, where each pair is composed of first an argument of the function and second its corresponding value.

An **exclusively extensional function** is an extensional function determined by a contingent function. Such functions relating empirical events are **extensional**, or **empirical**, **causations**, or high-valued correlations. Empirical causation has a fundamental weakness in that there is no clear demarcation between high correlations, which are empirical causations, and lower correlations, which are not; this is shown by that statistical problem of a correlation of 0.7, which is said to be too low to be significant but too high to be ignored.

A **nominal function** is an extensional function whose domain and/or range are nominal sets.

An **exclusively nominal function** is a name of a function that does not exist; among such are nominal functions whose domain, range, or both, are null.

8.   An old philosophical problem is the problem of universals. A universal is a word with plural references, such as 'Scotsman', as opposed to one with singular reference, such as James Watt (1736-1819) or James Clerk Maxwell (1831-1879). The problem of universals is: what is the nature of the meaning of universals? The medieval philosophers came up with three possible answers, but could not agree on which was correct. One answer was nominalism: the meaning of a universal is the word itself; as it was variously put, words are the counters of the mind, there is no thought without language, and all thought is silent speech. The second answer was conceptualism: a universal is a combination of word and abstract idea; the word is the universal and the abstract idea its meaning. The third answer was called realism, or Platonism: it affirmed concepts, but added that the abstract ideas were not simply abstract ideas, they were ideas of something real, such as Plato's ιδεοσ, or forms (207).

The problem was clarified somewhat by Descartes (213), who distinguished confused images from clear and distinct ideas. Confused images are concrete images in the imagination, while clear and distinct ideas are abstract ideas and are the content of thought. Descartes gave the examples of man, horse, and dog for confused images and mathematical ideas for clear and distinct ideas.

Clearly, the present work provides a definite answer to this problem: Platonism is correct. There are relations in the empirical world because we can perceive them; they are abstract entities because they have no concrete properties; we can think about them, so there are abstract ideas — ideal relations — in the mind; we can communicate them because they are bonded to words, to form concepts; these abstract ideas have explanatory value in so far as they are true — similar to relations in the noumenal world — and refer to these noumenal relations; and conceptual explanations explain empirical phenomena by describing their underlying, or noumenal, causes.

In ordinary language confused images and mathematical ideas are equally abstract and the words for them are equally universals. In this book the abstract is stipulatively restricted to thought and relational meanings, while confused images, which are concrete, are confined to the imagination and have extensional meaning. And language may describe either of these, or else purely nominal meaning, or else other kinds of meaning such as judgments, evaluations, memories, beliefs, questions, and commands.

9. When $R \Rightarrow E \Rightarrow N$, relational meaning has least generality and least arbitrariness, while nominal meaning has most of each and extensional meaning has a middling amount of each.

10. There are two important words which have nothing but nominal meaning. They might well be called cop-out words. One is the word infinity: when some people do not know the limits of something they declare it to have no limits, and so be infinite. The other is the word chance: when some people do not know the causes of something — they cannot explain it — they declare it to have no causes, and so be a matter of chance. Arguments for these two claims of the exclusively nominal meaning of infinity and chance will be found in Chapter 8.

11. As illustrations of paradox having nominal meaning only, we next analyse Russell's, Cantor's, and Grelling's paradoxes in terms of relational, extensional, and nominal

meanings. In each case we show that the paradox can have neither relational meaning nor extensional meaning, so has nominal meaning only.

Russell's paradox arises from the set, R, of all sets which are not self-membered, and its complement, R', the set of all sets which are self-membered; R is itself either self-membered or not self-membered; if R is self-membered then $R \in R$, and being a member of R, R is not self-membered; and if R is not self-membered then $R \notin R$, hence $R \in R'$, and being a member of R', R is self-membered. (Self-membership is supposedly justified by the examples of the set of all sets, which, being a set, is a member of itself; or the set of all abstractions, which is an abstraction and so a member of itself.)

Relationally, the relations of self-membership and non-self-membership are monadic[11] and so are exclusively nominal relations. Hence R and R' do not have intensions and so cannot be intensional sets.

Extensionally, since R and R' are not intensional sets, they must be contingent sets. But contingent sets also cannot be either self-membered or non-self-membered because there are no monadic extensional relations (since there are no one-membered ordered sets (87)), so every contingent set is neither self-membered nor non-self-membered. Also, neither R nor R' can be enumerated, hence cannot be defined extensionally. Hence neither R nor R' has extensional meaning.

So R and R' have nominal meaning only.

Cantor's paradox arises from the definition of a power set: the power set of an n-membered set S is the set of all the

---

[11] We could of course declare these monadic relations to exist but not multiply extravagantly, by denying them any uppers. But this would not explain anything, and so be denied by Occam's Razor. Or we could claim that they are ideal, because extravagant multiplication is possible in minds, but such ideal relations are of no significance because they have neither empirical nor noumenal reference.

subsets in S, including S itself, as an improper subset, and the null set; this power set is shown to have 2^n members, as follows. According to combination theory the number of ways in which k items may be selected from a set of n items, all different, is $n!/(k!(n-k)!$ so the number of k-membered subsets of an n-membered set is this number[12]. So the total number of subsets of an n-membered set is $\Sigma(n!/(k!(n-k)!))$ as k runs from 0 to n, inclusively. From the binomial theorem we know that $(a+b)^n$ is $\Sigma((a^{n-k} b^k n!)/(k!(n-k)!))$ as k goes from 0 to n, inclusively. So putting $a=b=1$ gives the total number of subsets as equal to 2^n. If now we define the set C as the set of all sets, and assume that C has c members, then C must have 2^c members, since every member of the power set of C is a set; thus C must have more members than it has.

If we look at this relationally, it is not immediately obvious whether the set of all intensional sets is itself an intensional set; but if it is not then the paradox cannot arise relationally. So let us assume that the set of all intensional sets is an intensional set, and call it C, which we can define as $C=\{\mathcal{E}\{\mathcal{E}x\}\}$, where x is any intension. The subsets of C, other than those defined by x, which are themselves intensional sets, take the form of $\{\mathcal{E}(\hat{A}\curlyvee\hat{B})\}$ or $\{\mathcal{E}(\hat{A}\wedge\hat{B})\}$, where $\hat{A}\in\{\mathcal{E}x\}$ and $\hat{B}\in\{\mathcal{E}x\}$ are any intensions, and provided that the disjunction is complete: $\{\mathcal{E}\hat{A}\}\cup\{\mathcal{E}\hat{B}\}=\{\mathcal{E}(\hat{A}\wedge\hat{B})\}$. If these conditions are met, $(\hat{A}\curlyvee\hat{B})$ and $(\hat{A}\wedge\hat{B})$ are intensions and so $(\hat{A}\curlyvee\hat{B})\in\{\mathcal{E}\hat{x}\}$ and $(\hat{A}\wedge\hat{B})\in\{\mathcal{E}x\}$. Consequently the number of subsets of C is equal to the number of members of C, and Cantor's paradox has no relational meaning.

Extensionally, the set of all extensional sets, by definition, consists of the union of two sets: the set of all intensional sets and the set of all contingent sets. The former is C, above, and already excluded leaving the set of exclusively extensional sets — the set of all contingent sets. This is either

---

[12] The symbol ! in mathematics means factorial: for all n, $n! = n\times(n-1)\times(n-2)\times...3\times2\times1$.

an intensional set or a contingent set. If it is an intensional set then it also belongs in *C*, and it has no intensional subsets, by definition, so it also is excluded. And if it is a contingent set of contingent sets then it can only be defined by enumeration and the enumeration of it is itself a contingent set of words, so the enumeration must contain itself as a proper part; this is possible only with infinite sets, and an infinite enumeration is impossible. So the extensional set of all extensional sets does not exist. So Cantor's paradox has no extensional meaning.

It follows that Cantor's paradox has nominal meaning only.

Kurt Grelling (1886-1942), a Jewish philosopher who was murdered by the Nazis at Auschwitz, defined the words homological and heterological, which produced his eponymous paradox. A word is homological if it describes itself: for example, the word short is short and the word English is English. And a word is heterological if it does not describe itself: for example, 'long' is not long, and 'eponymous' is not eponymous. Obviously, a word is heterological if and only if it is not homological. If now we ask whether the word heterological is heterological or homological we get a paradox. Because if 'heterological' is heterological then it describes itself, in which case it is homological and so not heterological. But if 'heterological' is not heterological then it does not describe itself, in which case it is heterological.

Relationally, description is a relation between a word or sentence and what it describes, and this relation is dyadic; call it D. To say that a word has relational meaning is to say that the relation D exists between that word and that meaning. The word homological requires D to be monadic: that is, D is self-referential. But D cannot be monadic because relationally there are no monadic relations except trivially, so the word homological has no relational meaning. So every word that has relational meaning is heterological — provided that heterological has relational meaning. But 'heterological' means the absence of D, and the absence of a relation has no relational meaning. So the paradox does not arise relationally.

Extensionally, we do not know if the words heterological and homological have extensional meaning, so let us assume that they do. This means that there are the two sets, of all heterological words and of all homological words. But these sets must be contingent because the words heterological and homological have no relational meaning, hence cannot be intensions. So the membership of these sets is determined by enumeration, hence according to the paradox the word heterological is either in both enumerations or else in neither, in which case it does not have extensional meaning — to say nothing of the fact that both sets are probably too large to enumerate and so do not exist.

It follows that Grelling's paradox has nominal meaning only.

If you have a lingering doubt about this, on the grounds that some words do *in fact* describe themselves, so that the word homological is in fact meaningful, then consider that 'homological' means self-descriptive and self-description is a monadic relation, like self-similarity, self-membership, and self-verification[13], and so 'homological' has purely nominal meaning because there are no such monadic relations — *except in language.* All paradoxes exist in language, and only in language, and that includes paradoxes about language. Or, to be more precise, they exist only in language or else trivially in minds.

Table 6.1 summarises much of our examples of the three kinds of meaning. In each of the three columns the entry gives the most significant feature of that kind of meaning.

Finally, what is the value of distinguishing these three kinds of meaning, relational, extensional, and nominal? First,

---

[13] For example, what I write three times is true, because what I write three times is true, what I write three times is true.

to discover purely nominal meaning is to discover, and eliminate, nonsense, babble, and the purely rhetorical. Secondly, to distinguish the extensional from the relational has the more technical, but important, value of denying the tradition of basing mathematics on set theory in favour of basing it on relations and relational meaning. And thirdly, we shall discover, in Chapters 8 and 11, that the noumenal has only relational meaning; so if you seek truth in explanations then concentrate on relations.

6. Three Kinds of Meaning

	Relational meaning	Extensional meaning	Nominal meaning
Word or description	Relations or properties thereof	Intensional or contingent sets	Non-referential names
Characteristic	Axiom generosity	Consistency	Contradiction
Necessity	Singular possibility	Universality	Tautology
Analyticity	Predicate is contained in subject	Subject class is part of the predicate class	Denial produces a contradiction
Set membership	Necessitated	Contingent	Non-referential
Set defined by	Intension	Enumeration	Declaration
Set	Intensional	Extensional	Non-referential
Relation	Primitive	Ordered set	Polyadic predicate
Sentential connectives	Defined by similarity	Defined by identity	Defined by truth-tables
Synthetic truth	Similarity to reality	Similarity to reality	Correct usage
Negation	Dissimilarity to reality or else set complementarity	Set complementarity or non-membership	Truth-functional
Conjunction	Coupling	Intersection	Truth-functional
Disjunction	Commonality	Union	Truth-functional
Implication	Necessity	Subset	Truth-functional
Equivalence	Similarity	Set identity	Truth-functional
Number	Adicity	Cartesian product	Non-referential
Validity	Necessity	Subset	Tautology
Generality	Minimum	Medium	Maximum
Arbitrariness	Minimum	Medium	Maximum
Axiom generosity	Yes	No	No
Contradictions	No	No	Yes

Table 6.1.

Chapter 7. Explanation.

Quality of explanation is a criterion for believing in what has noumenal existence (28). As such it is an answer to Hume's contention that 'tis vain to speculate (28), since the explanatory value of a concept or proposition is the probability[14] of it having noumenal reference. Does the existence of mermaids explain anything? No; so there are no noumenal mermaids. Does the standard model in quantum mechanics explain a lot? Yes; so it almost certainly has noumenal reference.

So we next examine quality of explanation. There are criteria of good explanation: they apply to all explanations — explanations in the form of lunatic fringe explanations, myths, common sense, theology, metaphysics, and science. There are at least twelve such criteria.

1, 2.   The first two are falsifying criteria: if an explanation contains or entails a *contradiction* then it is false, and if it *contradicts established empirical fact* then it is false. It is widely believed by scientists that these are the criteria that distinguish science from all other kinds of explanation: if a theory cannot in principle be falsified then it is unscientific. This claim, due to Karl Popper (1902-1994), is a legacy from the school of philosophy called logical-positivism, which wanted to claim that only logic and empirical data were meaningful: all other theories, such as metaphysics and theology, were thus meaningless. Their original definition of meaningfulness was that only empirically verifiable, or logically valid, statements, were meaningful; but this was self-refuting, since it cannot be verified, so they fell back on falsification: only statements that can in principle be falsified are meaningful. However this statement is also self-refuting.

---

[14] This probability is the strength of belief, of informed people, in the truth of the proposition, not the probability that is the reciprocal of degree of possibility, nor statistical relative frequency.

7. Explanation

But this did not quite express their views, because they were essentially empiricists, believing that only empirical data, and logically valid statements about them, were meaningful; or, more succinctly, only empirical science and logic are meaningful. But this made theoretical science meaningless, and they could not deny such a large part of science. So, after several failed attempts to prove that theoretical science is somehow really empirical, logical-positivism has been largely abandoned. Consequently, although falsification remains an important criterion, it is not the only one.

All the other criteria might be called verifying criteria; they verify in the sense of making an explanation more probable, meaning more generally believed by suitably informed people.

3, 4.   The first two of the verifying criteria are the complementary *density of detail* in an explanation, and the *scope* of an explanation. Jointly, these show how much empirical fact an explanation explains, and, obviously, the more it explains the better it is as an explanation. An example of an explanation with large density of detail is the standard model in quantum mechanics; an example of an explanation with large scope is evolution, which explains the huge variety of living things, and of the fossil record. An explanation with maximum possible scope is the explanation that everything happens as it does because it is the will of God; but since this explanation has zero density of detail, it is useless as an explanation. Generally, scope is like a forest and density of detail is like trees: the bigger the forest the greater the scope, and the more detail in the trees — such as details of branches, twigs and leaves — the greater the density of detail. Another helpful metaphor is the resolution of the lens of a camera: if a camera has a wide-angle lens the photograph has large scope, and a telescopic lens produces density of detail. A measure of each, multiplied together, gives the resolving power of the lens, which is like the quality of the explanation.

5.   *Coherence* is another criterion: a theory is coherent — it has coherence truth — if it is rich in internal necessities, or, equivalently, lacking in internal arbitrariness or contingency. The inability of the standard model to explain the cosmic coincidences (160), or the collapse of the wave-function, are examples of lack of coherence.

6.   Another criterion is *beauty* — thanks largely to Paul Dirac (1902-1984). Scientists are usually a little coy about beauty — it seems too subjective — and use the terms elegant or pretty instead; but beauty is the proper term to use, philosophically. If two explanations are otherwise equal, the more beautiful one is to be preferred. And if a theory is ugly it is probably false.

7.   *Simplicity* is another criterion. If two explanations are otherwise equal the simpler one is to be preferred. And if a theory is complicated it is probably false — although, of course, if it is over-simplified it is also probably false.

8.   *Harmony* with other accepted explanations is also regarded as a criterion, particularly by physicists and mathematicians. Harmony is exhibited when two theories are combined into one, as with Descartes' (1596-1650) unification of algebra and geometry in co-ordinate geometry; as with Maxwell's (1831-1879) equations which combined, electricity, magnetism, and optics; and as with physical chemistry, which united quantum mechanics and chemistry. An example of a major absence of harmony is that between general relativity and quantum mechanics.

9.   *Symmetries* are a criterion — in physics, at least. Symmetries are related to conservation by Emmy Noether's (1882-1935) famous theorem and are expressed mathematically in group theory. These days principles of

conservation such as conservation of energy, momentum, and electric charge are explained by means of group theory.

10.   *Occam's Razor* — do not invent more theoretical entities than are needed to explain the empirical facts — is an important criterion. Or rather, it is part of one, the other part being its converse: do not *reduce* entities beyond necessity. If, for example, a noumenal world is needed in order to explain explanation by means of imperceptible causes, then do not deny it.

11.   *Cascading emergence* of theorems in an explanation is a significant criterion because of its connection with the Grand Structure and with the next criterion.

12.   Last, the criterion of overwhelming importance is *theoretical prediction of empirical novelty.* Mathematical theories in science can do this, and we need to know how they do it. (Sometimes empirical novelty is predicted by interpolation, as William Harvey (1578-1657) predicted the existence of capillaries between arteries and veins in order to explain circulation of the blood, and Mendeliev (1834-1907) predicted new chemical elements from gaps in his periodic table; but these are rare.) Three best known examples of theoretical prediction of empirical novelty are the prediction, by Maxwell's equations, of a vast range of electromagnetic frequencies beyond those of visible light and radiant heat; the prediction of nuclear energy from Einstein's (1879-1955) famous equation, $e = mc^2$; and the prediction of the Higgs boson, by Peter Higgs (1929-) and others, in the Standard Model and verified by the Large Hadron Collider in 2012. Modern mathematical scientific theories make predictions of novelty successfully and often, and such success is convincing evidence for the truth of these theories.

      So how do these theories do this? There is only one good answer. Predictions of empirical novelty are logically necessitated by the mathematical theory, and the predictions come true because the novel empirical phenomena are causally necessitated within the underlying noumenal world; and the logical necessities are analogues of the causal necessities. If the predictions were not logically necessitated then they would be mere guesswork; and if the novel phenomena were not causally necessitated then they would happen only by chance. Guesswork and chance could conceivably combine successfully on extremely rare occasions, but the sheer quantity of successful predictions of empirical novelty in the history of physics, astronomy, cosmology, and chemistry rules this out — to say nothing of the fact that there is no relational chance (113). So we can say with assurance that mathematical theoretical science contains logical necessities, and the underlying noumenal world contains causal necessities, some of which latter cause empirical phenomena. Not only that, but the antecedents of the logical necessities must describe essential features of the causes, and the consequents of the logical necessities must describe essential features of the effects — that is, of the empirical novelties. So, in short, mathematical theories predict empirical novelties, successfully and often, only because they are true. They are true in the sense of correctly describing, mathematically, essential features of the noumenal world. This is so important in philosophy of science, as a criterion of good theory that it is worth going into greater detail, with a specific example.

Our example is illustrated with a pattern diagram in Fig. 7.1. It begins with a noumenal cause, A_1, and its noumenal effects, L_1 and T_1. We may suppose that A_1 is a noumenal atmospheric electric discharge, L_1 is a noumenal flash of lightning and T_1 is a noumenal clap of thunder. That is, L_1 is a burst of electromagnetic radiation in the visible spectrum and T_1 is a burst of atmospheric acoustical vibration, both of these radiating outwards in all directions as waves. Each of these noumenal effects, L_1 and T_1, noumenally causes an image of itself, L_2 and T_2, in the empirical world. Each of

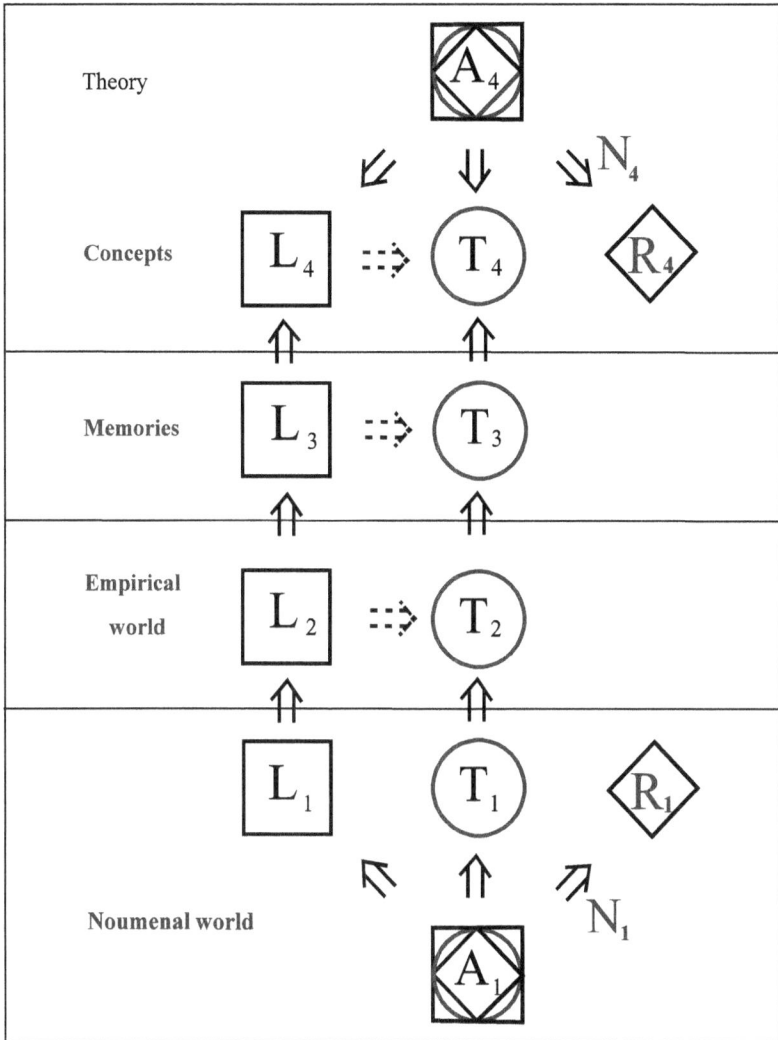

Fig. 7.1.

these images is empirically observed by a subject in the empirical world, as empirical lightning and empirical thunder. Each such empirical phenomenon causes an image of itself — a memory — in that subject's mind. Each such image — remembered lightning and remembered thunder, L_3 and T_3 —

then causes the formation of concepts, L_4 and T_4, by a process that for now we simply take for granted (see Chapter 11), where a concept is a combination of a word or symbol and the abstract idea that is its meaning. The solid arrows in Fig. 7.1, represent necessities. Thus A_1 noumenally, causally, necessitates, L_1 and T_1; L_1 similarly necessitates L_2, which necessitates L_3 which necessitates L_4; and T_1 similarly necessitates T_2, T_3, and T_4, transitively. (It is of course assumed that all other necessary conditions are present; for example, L_1 would not necessitate L_2 if there was no person present to perceive L_2, and L_3 would not necessitate L_4 if the person concerned did not think about what he or she empirically perceived.) So L_4 and T_4 are concepts of lightning and thunder. Notice that L_1, L_2, L_3, and L_4 are all qualitatively different from each other and so are numerically distinct: it is impossible for any two of them to be identical, one and the same, because of Q.Q.: qualitative difference entails quantitative difference (70); equally so for T_1, T_2, T_3, and T_4. The broken arrows represent correlation — which may be called **empirical**, or **extensional**, **causation**. In an empirical world this is simply constant conjunction of contiguous and successive events, as David Hume put it (that is, sets of similar pairs which are terms of temporal joins) and in an empirical mind it is memories of these, and association of ideas. Thus the correlation between L_2 and T_2 is an instance of the trivial empirical law that lightning empirically causes thunder; and a remembering or thought of lightning, L_3 or L_4, is followed by a remembering or thought of thunder, T_3 or T_4, by association. Empirical causation (111) is of course a misnomer at best: the common belief that empirical lightning causes empirical thunder is false — empirical lightning and thunder are both caused by a noumenal atmospheric electric discharge, and they are correlated because of this common cause.

If the person in whose mind all this is occurring has some basic physics, he or she will have a concept, A_4, of an atmospheric electric discharge, and of how this causes thunder and lightning. The solid arrows between A_4 and L_4, and between A_4 and T_4, represent the necessities within the theory that contains A_4, L_4, and T_4 such that within this theory A_4 necessitates L_4 and T_4, logically. In other words, the arrows from A_4 represent logical necessity, as opposed to the causal necessity of all the other heavy arrows lower in the diagram. Thus L_4 and T_4 can be deduced from A_4, and so can their correlation.

If the theory in the mind of Fig. 7.1 is true, it is because it copies, as accurately as possible, a state of affairs in the noumenal world: A_4 accurately copies A_1, L_4 accurately copies L_1, T_4 accurately copies T_1, and the logical necessities between A_4 and L_4 and between A_4 and T_4, accurately copy the causal necessities between A_1 and L_1 and between A_1 and T_1. Thus the empirical law that lightning causes thunder is explained in the sense of describing the underlying causes of lightning and thunder.

It is such logical necessities that enable theoretical predictions of empirical novelty to be made, and such causal necessities that make these predictions come true. Suppose now that A_1 regularly causes R_1, as shown by the arrow N_1, but that R_1 is not perceived as R_2 because R_1 is outside the range of human perception, so is noumenally imperceptible. In the present example, R_1 is electromagnetic radiation outside the visible spectrum and radiant heat. (The visible spectrum of electromagnetic radiation is noumenally visible, but not empirically visible; it is the cause of the empirical visible spectrum — the colours of the rainbow.) If the theory in which A_4 necessitates L_4 and T_4 is true, then A_4 will logically, mathematically, necessitate R_4, as shown by the arrow N_4. When the scientist deduces R_4 from the theory, he is predicting empirical novelty. In this particular case he would be predicting radio noise. An experimental scientist should then be able to figure out what is needed — some experimental apparatus — to make R_1 noumenally perceptible. When she has built this apparatus, a radio receiver, and set it up during an electric storm — a situation such that her empirical world will contain instances of L_2 and T_2 — then she will hear radio noise, R_2, correlated with every flash of lightning, L_2, by means of it. The prediction of novelty will then have come true. It comes true because of the causal necessity, N_1.

As far as I know no scientist ever made this particular prediction of novelty, but it could easily have been made and it serves well as an illustration even if it should be historically inaccurate; and in fact it could well have been made by Heinrich Hertz (1857-1894), who first discovered such electromagnetic waves, as predicted by Maxwell.

Thus the possibility of prediction of novel empirical facts by theoretical deductions is explained. Minimum conditions for this are a four-way similarity between (i) part of the structure of the noumenal world, (ii) part of the structure of the empirical world, (iii) scientific laws, and (iv) theories that are both mental constructs out of (iii) and mental reconstructs of (i). Without the noumenal world, theoretical prediction of empirical novelty is inexplicable, because there are no necessities in empirical causation — correlation — to correspond to the logical necessity in the theory. In other words, if prediction of novelty is to be explained then believing in the existence of a noumenal world that underlies the empirical world and is described by theoretical science is necessary; and, more particularly, a world containing causal necessities described by the mathematical necessities in theoretical science is necessary. Without the mathematical necessities, such as N_4, the prediction of novelty would be mere guesswork, and without the causal necessities, such as N_1, the success of the prediction would be pure chance.

There is one further point about theoretical prediction of empirical novelty, which is quite important. Novelties, as intrinsic properties of relations, *emerge*. The Grand Structure shows this at every level. Only relations emerge. And our main language of relations is mathematics. So that is why in general only mathematical theories predict empirical novelties.

All these twelve criteria of explanation, when applied to the kind of speculation that Hume denounced, provide a value to an explanation: the greater the explanatory value the greater the probability that the speculation truly describes the noumenal.

The relationship between the noumenal world and the empirical world is explained in Chapter 11.

A general remark about the verifying criteria of explanation with regard to philosophy of science is to compare philosophy of science to philosophy of law and justice. In a criminal court a decision needs to be made about something that cannot be perceived: a crime that occurred in the past, for which, in the present, there is only evidence; evidence for and against the crime and the accused criminal. Something imperceptible, and empirical evidence for or against that something, are quite distinct, because of Q.Q. (70): the evidence is empirical, and the imperceptible something is not empirical. Equally, a theory about the noumenal world and evidence for or against it are quite distinct: the evidence is empirical and the noumenal is not — by definition of the noumenal (7). The court requires a conclusion based on what any reasonable person would believe, and this conclusion is reached by a jury of twelve supposedly reasonable people. So a theory is accepted, or verified, when it is believed by a majority of scientists — an informal jury of supposedly reasonable people. In the language of science a theory becomes more probable the more it conforms to the above verifying criteria of explanation. This probability is neither mathematical probability nor statistical relative frequency, it is the strength of rational belief of the jury. Further justification of these criteria is given in the next chapter.

Finally, the present book is an explanation, of mathematics, science, and mind. Anyone who wishes to criticise it, positively or negatively, is invited to use these twelve criteria in their evaluation.

Chapter 8. Possible Worlds.

In this chapter we investigate the variety of possible worlds, in order to try to discover which of them best describes the noumenal world.

A **relational possible world**, or **possible world** for short, consists entirely of ideal relations, which is to say that it exists only in the mind of a thinker. (We include computers here, in so far as they emulate mental operations such as making comparisons.) A possible world is an idea of a possible noumenal world, not of a possible empirical world. One of all possible worlds is an idea of the actual noumenal world; only one possible world can be this because there is only one noumenal world, since the noumenal world is defined as the totality of noumena — the non-empirical things considered in theoretical science. So among every possible world we only consider those which may be a candidate for being the one synthetically true (52) description of the noumenal world. Since the noumenal world is complete and includes the Grand Structure (8), so must every possible world be complete and possess a grand structure. This means that each possible world has a series of levels, emerging from an axiom level and ending in a single top relation; and this top relation has no uppers and has an emergent property unique to that world. In more detail: we earlier made three definitions: a structure is a series of levels of relations relating relations, having a top level of a single relation; a whole is a structure that has a novel property in its top relation; and a possible world is a whole with a top relation which is not a term of a higher level relation. It is this fact of a top relation having no uppers that makes a possible world complete; it is what makes it a possible world rather than simply a whole. Because it is complete it follows that no two possible worlds can co-exist. Because if they did co-exist then their top relations could be related and the two worlds would then become one possible world, in which case the original two would neither have been complete nor possible worlds. (To say that their top relations *could* be related means, logically, that a relation having them as terms is *possible*, which gives it mathematical existence.) In fact, if anything whatever existed outside a possible world, then that thing could be related to the top relation of that possible world — e.g. the two are dissimilar — thereby negating its status as a possible world by making it incomplete.

The distinction between the noumenal world, on the one hand, and the set of all possible worlds on the other, is that we believe that the noumenal world has actual existence — because of the success of the sciences — while possible worlds have only ideal existence, they exist only in minds of thinkers. There is only one noumenal world, but we can think of many possible worlds, each a candidate for being a true description of the noumenal world. So we have to distinguish two kinds of ideal existence: **worldly existence** and **exotic existence**. A possible world has worldly existence in so far as it is a candidate for describing the noumenal world; as such, two possible worlds cannot co-exist. But we need to compare various possible worlds, and when we do this the comparison has exotic existence; as such two possible worlds *can* co-exist. So we can define two further kinds of ideal relations: **worldly relations** are ideal relations *within* a possible noumenal world — that is, an ideal possible world — and **exotic relations** are ideal relations *between* possible noumenal worlds. Thus there are no exotic noumenal relations but there are exotic ideal relations.

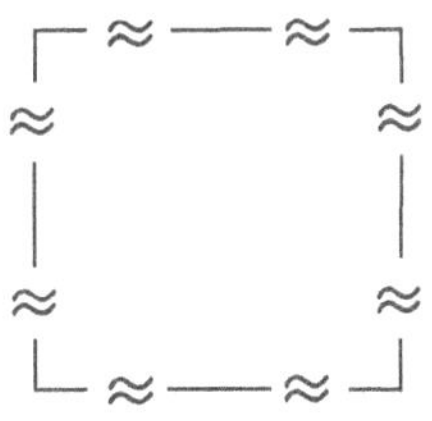

Fig. 8.1.

In beginning a comparison of possible worlds we first look at the simplest possible world (Fig. 8.1). For this we assume two kinds of separator at the axiom level: an atomic (that is, uncuttable) length, called a **linear separator**, and an atomic angle which is a right angle and called an **angle separator**. The **simplest possible world**, or atomic square (100), has a level-2 top relation and terms consisting of four right angle separators, four linear separators, and eight dissimilarities, such that each linear separator separates two angle separators, each of which separates two linear separators; and between each pair of linear separator and angle separator there is a relation of dissimilarity; out of this emerges the top relation, called a square, having the emergent property of an area.

In a larger possible world, with an axiom level also containing linear separators and angle separators, there might be many squares; each would be a whole in that world, an atomic square. They might be thought of as **quasi-separators** because squares could separate squares; but they would not be true separators because chains of them could end (31). Atomic squares would also be compoundable relations (38) because they could combine to form larger areas in level-3; these would be determined by their boundaries — all the dissimilarities between square and non-square. Level-3 could also have an emergent relation with a novel property: six atomic squares, twelve angle separators, and twenty four dissimilarities could form an atomic cube, having the novel property of a volume. ('Novel', remember, in this context means that it does not emerge at any lower level.) Thus increasing the number of separators in the axiom level leads to rapid increases of emergents.

If we define the **population** of a level as the number of *instances* of all the relations in that level, and the **variety** of a level as the number of *kinds* of relations in that level as well as in all lower levels, then the above illustrations show two important things: as you go to higher and higher levels the population decreases and the variety increases. The population decreases because emergent relations must each have at least two terms, and the variety increases because of emergence of novelties. This means that the top level of a possible world must have maximum variety, and a minimum population of one, and the axiom level of that possible world must have minimum variety and the maximum population; these minima and maxima are, of course, relative to that possible world. We see this with what we know, scientifically, of the noumenal world: the Grand Structure (8) is a strong indication that the noumenal world is indeed a structure of relations relating relations, for many levels, and partially described by one of the possible worlds that are being investigated in this chapter. Secondly, the fact that the variety of possible molecules is greater than the variety of atoms illustrates the increase of variety and the decrease of population with increase of level. Thirdly, the Grand Structure is the focus of all the theoretical sciences: every branch of theoretical science concentrates on some portion of the Grand Structure, so that if there were no Grand Structure in the noumenal world all of theoretical science would be fiction.

A particular kind of property or relation emerges at the same emergent level in whatever possible world that it in fact emerges. The reason for this is that if it emerges only at a certain level, and never at a lower level, it must be because emergence at a lower level is impossible; and this impossibility applies in all possible worlds. For example, life emerges out of complex structures of molecules, so we cannot expect atoms or simple molecules to be alive, in any possible world that has these; and intelligence emerges out of brain, so we cannot expect bacteria to be intelligent, in any possible world that has bacteria.

We earlier (37) developed a measure of the quality of a possible world: this is the total amount of emergent hekergy in it, divided by the number of separators in its axiom level: a reasonable definition, since a possible world consists entirely of relations, separators are not emergent and do not have hekergy (because they do not emerge from emergent configurations), and the hekergy of an emergent relation is the absolute value of that relation. Because the population in each level is fewer than the amount in the next lower level, a possible world is somewhat analogous to a right cone. The amount of emergent hekergy is analogous to the volume of the cone, and the number of axiom relations — axiom level, or level-1, separators — is analogous to the area of the base of the cone. Since the volume of the cone is $\pi r^2 h/3$, and the area of the base is πr^2, where r is the radius of the base and h the height of the cone, the quality, q, of a possible world is proportional to h, or the emergent level of the top relation of that world, T: $\pi r^2 h/3\pi r^2 = h/3$. So we may say that the quality of a possible world may be represented by the level of its top relation, T, symbolised by #T: $q=k.\#T$, where k is a constant. Because every possible world has a top level, #T, possible worlds may be ranked by their quality. In fact, since every possible world has a number, #T, the set of all possible worlds forms a partially ordered set (a poset) and this fact requires that there is at least one possible world of maximum quality — at least one best of all possible worlds.

We next return to the concept of loops (31). The two key features of these loops are that what is being looped is possibility, which is mathematical existence and one of the three properties in the minim; and the links are necessities: inherence, emergence and demergence. So the loops are necessary chains of mathematical existence, which is to say chains of singular possibilities of possibility.

Besides mathematical existence — possibility — which is possessed by every possible noumenal world, there is also actual existence, or **actuality**, which is the existence possessed by both the empirical world and the noumenal world. The one possible ideal noumenal world which truly describes the noumenal world, does so because it has the idea of actuality, which is **ideal actuality**. Ideal actuality is the idea of intrinsic necessity, which is intrinsic singular possibility — as opposed to mathematical existence, which is intrinsic possibility. That is, actuality is a special case of possibility. Remember that the possibility in the minim, mathematical existence, is plural possibility, contingency (44); when this is singular possibility it is actual existence.

Actuality, in the first place, is empirical existence, which is best defined, stipulatively and personally, by Descartes' *cogito ergo sum*: I am conscious therefore I exist; we all know it empirically. The existence of *me* is actual existence. An important fact about empirical actuality is that actual relations cannot exist without actual terms, so that an actual relation demerges actuality; and equally a set of actual relations with an actual emergence-configuration necessitate the emergence of a set of properties and hence of an actual relation. In short, actuality demerges, emerges, and inheres actuality. This means that if one relation in a possible world has actuality then the loop possibility of that world becomes loop necessity: if one relation is actual then the whole world is actual. Recall that the noumenal world is postulated in the first place because of the need to explain; and that explanation is causal — to describe causes is to explain their effects; and also that causes are underlying, non-empirical, hence noumenal. So empirical actuality is explained by describing **noumenal actuality** because, as we shall see in Chapter 11, noumenal actuality is the cause of empirical actuality.

Noumenal actuality is intrinsic necessity, as opposed to mathematical existence, which is intrinsic possibility. It is this noumenal actuality that is meant when it is said that only one noumenal world can exist: that is, only one noumenal world can be actual. The essential nature of noumenal actuality is a change in the minim of the relations of the noumenal world: the intrinsic property of possibility in the minim is changed to necessity: mathematical, or logical, or possible, existence is changed to actual, or necessary, existence; which is to say that plural possibility is changed to singular possibility, loop possibility is changed to **loop necessity**, and loop mathematical existence is changed to loop actuality.

We have to distinguish the noumenal world and the ideal possible world that truly describes it, and distinguish noumenal actuality and ideal actuality, because although they are similar — the truth of the ideal is similarity truth — they are none the less distinct: one is real and the other is ideal. But to constantly distinguish them can get tedious; so, unless the context requires otherwise, in what follows they will each be treated as one (the identity error!) even though it must be understood that they are two.

Noumenal actuality is symbolised by ⊡: the symbol □ is the standard logical symbol for necessity and here the dot inside signifies that it is intrinsic. Similarly, possibility is symbolised by ◊ and intrinsic possibility by ◈. We shall see in Chapter 11 that empirical actuality is a special case of noumenal actuality.

Noumenal actuality originates with **necessary actuality**, ⊡, which is a property that emerges in a relation, S. That is, S has the property of possibility in its minim changed from plural possibility to singular possibility; S no longer has mathematical existence, it has actual existence: that is, ◈S becomes ⊡S. And S, in being actual, loop necessitates the actuality of the world that contains it. So a possible world that possesses S becomes thereby an actual world.

S has to be a relation that is unique among all possible worlds. If it was not unique among all possible worlds there would be more than one actual world, which is impossible. There is only one way in which a relation can be unique among all possible worlds and that is to be the top relation of a possible world that has a higher top level than any other possible world. We will call this unique top relation T. So S=T. The possible world that has T as a top relation has to be one of the best of all possible worlds, by definition of the quality of a possible world; and there can only be one such best, because T is unique among all possible worlds. This best possible world is symbolised by G. The terms of T have an emergence-configuration, E, that extrinsically necessitates, $\dot{\square}$, the novel property of ⊡, so T has necessary actuality: E$\dot{\square}$(⊡T) or E⇑(⊡T) and T thereby necessitates, by demergence and emergence, the actuality, ⊡, of the rest of the world of which it is the top relation. Or, in other words, the necessary actuality in T, ⊡, changes the loop possibility of its world to loop necessity; it changes a possible world into an actual world; it changes the property of possibility, ◊, in the minim of all its world's relations, which is plural possibility, to necessity, ⊡, which is singular possibility. The demergence and emergence by which T does this are extrinsic necessities, ⊡̇, or ⇓ and ⇑; they constitute a loop necessity. Without the necessary actuality that emerges with T, nothing would be actual; but once the unique ⊡T emerges — ⊡̇ ⊡T — the entire possible world that has T as its top relation becomes actual, ⊡G.

We have just proved that the one and only best of all possible worlds actually exists, necessarily, because it is the best. This may be spelled out in greater detail.

1.   If two possible worlds were actual then the two could be related into one, in which case neither would be complete, hence not a possible world. So *at most one possible world can be actual.* (Italicised statements here and in the next three paragraphs identify five crucial conclusions.)

2.   Some relations are actual because there are relations in the actual empirical world, and empirical actuality is a special case of noumenal actuality because empirical phenomena are actual and are caused by actual noumena. Therefore the noumenal world is actual. So *at least one possible world is actual.*

3.   If more than one relation possessing necessary actuality, ⊡, were actual, in different possible worlds then two or more possible worlds would be actual, which is impossible — by 1, above. So *at most one possible world possessing a relation possessing necessary actuality can be actual.* And if two relations possessing necessary actuality were actual in one possible world then one of them would have to have an emergent level lower than the top level of that world (see 5 below), in which case such a relation could emerge at that

level in another possible world of equal or higher top level, and make that second possible world actual — which is impossible. So *a relation possessing necessary actuality must be the top relation of a possible world.*

4.   The one possible world that is actual has nothing actual outside of it, so its actuality cannot be externally caused; and it cannot be actual by chance because there is no relational chance (113). So *it must be internally actualised by possessing a relation that has necessary actuality.*

5.   The only way in which all five of these italicised conclusions can be met is that (i) the actual possible world has a top relation having necessary actuality; (ii) the top level of this possible world is higher than the top level of every other possible world and so unique among all possible worlds; and (iii) this top level is the emergent level of actuality.

6.   Thus the world with this unique highest level is the best of all possible worlds, G; and its top relation, T, is the one and only relation, among all possible relations, possessing actuality that is both emergent and intrinsic. In other words, G is the best of all possible worlds and is actual because it is the best. That is, it is the best because it has maximum hekergy because it has the highest possible top level, which means that it has maximum quality (37); and it is actual because this top level is the unique emergent level of actuality.

7.   Finally, the noumenal world is defined as the totality of noumena, so there can only be one noumenal world. And because empirical phenomena are actual and are caused by noumena, noumena must be actual. And because G is the ideal possible world that both truly describes the noumenal world and is the totality of worldly actuality, the actual noumenal world must be the best of all possible noumenal worlds. So the

noumenal world actually exists necessarily because it is the best of all possibiles.

This argument is a version of the ontological argument, which originated with St. Anselm (1033-1109), who used the phrases "that, than which nothing greater can be conceived" and "necessary existence". His argument essentially was that that, than which nothing greater can be conceived exists necessarily because if it did not exist it would be less than that, than which nothing greater can be conceived. St. Anselm was concerned to prove the existence of God, and the noumenal world is one of the meanings of the word God (see Chapter 13). Immanuel Kant (1724-1804) invented the name ontological argument and popularised the main objection to it: that greatness of conception is greatness of predicates, and existence is not a predicate, so the non-existence of that, than which nothing greater can be conceived, does not diminish it. The present version of it shows Kant to be wrong: a predicate of a relation is a property of that relation and both intrinsic possibility (mathematical existence) and intrinsic necessity (actual existence) are properties — predicates — of relations. Furthermore, actual existence has higher hekergy then mathematical existence, because singular possibility has higher hekergy than plural possibility; this is because hekergy is defined as $\ln.t/e$ and singular possibility means that $e=1$ while plural possibility means that $e>1$. So the quality of G would diminish if it did not actually exist.

The best of all possibles is a deterministic world, a world in which there are no chance events. Although this appears to contradict quantum mechanics and statistical mechanics, this is not necessarily so. We saw in Chapter 1 (44) that probability is the reciprocal of possibility: of the three modalities — contingency, necessity, and impossibility, which are respectively plural, singular, and zero possibility — if a possibility is n and $n>1$ then n defines a plurality and a contingency, and *1/n=p* is a probability. It is this kind of probability that occurs in quantum mechanics and statistical mechanics: a modal probability. As such it is horizontal, like a causation, but uncaused. But although uncaused, such contingencies are none the less necessitated, in a deterministic world, by either downward or upward necessitation — by demergence or emergence. Thus the best of all possible worlds may contain contingencies (but no radical contingencies, or undetermined events) while being fully determined. The contingencies may be thought of as ignorance-placeholders.

This conclusion is supported by a plausible conjecture that comes out of the ontological argument: namely, that hekergy is conserved. Hekergy has to be a maximum in the best possible world, since 'the best' is defined in terms of quantity of hekergy, and this would also mean a maximum over time. Combining this conjecture with the denial of chance would then mean that all of the supposedly chance positive mutations and gene combinations in evolution — biological hekergy increases — would cancel out all the supposedly chance increases of entropy in thermodynamics — entropy being negative hekergy — and supposedly chance collapses of the wave-function in quantum mechanics. (And indeed, in quantum mechanics entropy — information — is conserved.) In other words, evolution is a hekergy increasing process equal and opposite to, and upwardly necessitated by, the hekergy decreasing processes of thermodynamics. (The denial of chance is further discussed in Chapter 10 (153).)

There is also nothing infinite in the best of all possible worlds. It might be argued that there is no difficulty in defining an axiom level much greater than that of G, and that this would thereby produce a top level higher than that of G, thereby falsifying the ontological argument. To show the fallacy of this, consider a parallel case: we saw in Chapter 5 that if there is no relational meaning to infinity it follows that there must be a greatest relational cardinal number, *g*. We can then define a number, *g+n*, where *n* is any number, and

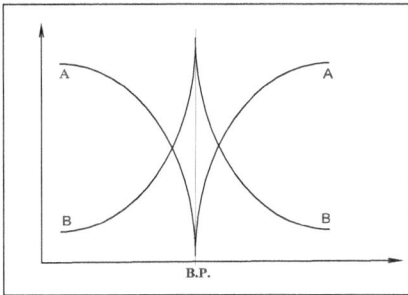

Fig. 8.2.

thereby "prove" that g is not the greatest cardinal. However, to define something does not mean that it exists, and if it does not exist then the definition has nominal meaning only — it has no relational meaning; also, such a definition assumes the principle of closure on addition, which requires that the sum of any two numbers is a number — a principle that is not necessarily true. In other words, g is the greatest *relational* number and any $g+n$ does not exist except nominally. Similarly, G is the relational possible world with the highest top level; anything else that is claimed to be higher has only nominal meaning.

Understanding of this may be helped by consideration of the fact that the ontological argument is the ultimate stationary principle, as shown in Fig. 8.2; in this figure the x-axis represents the number of axiom separators in the axiom level of various possible worlds and the y-axis shows both the number of resulting possible worlds (AA) and their quality (BB). If the number of separators in an axiom level be increased from small to very large then the resulting variety of possible worlds decreases (AA), and will at first have increasing emergent levels of their top relations (BB); and at low levels of top relation the variety of possible worlds will be large (AA), but this variety decreases with increasing top level. So at some point B.P. a maximum top level must be reached, after which such top emergent levels will again decrease in level, and variety will again increase. And the increasing variety is ideal and so cannot increase to infinity because there is no ideal infinity, only a word 'infinity' that has nominal meaning but no reference. There is thus a maximum top level, of variety one. The cusps of each curve are of course at the same point, B.P., on the x-axis.

We also note that a true explanation is true (129) by correctly describing features of the noumenal world. Since the actual noumenal world is the best of all possibles, true explanations cannot contain contradictions or be contrary to empirical reality (that is, to non-illusory empirical data); they must be of high hekergy hence be simple and beautiful; they must harmonise with other true explanations and be coherent; and they must be necessitous, hence able to predict empirical novelty. These are the criteria of good explanation, and by these criteria all of the theoretical sciences jointly and (largely) truly describe the Grand Structure.

There seems to be a threat of paradox in the idea of separator loops: what would happen if a nihilist, wanting to destroy the Universe, went down to the axiom level of the noumenal world and destroyed one link in a closed loop of separators? The entire noumenal world would immediately cease to exist. This is similar to the grandfather paradox: what would happen if a time-traveller went back in time and killed her grandfather before he had any children? In each case the destroyer would self-destroy and so could not destroy. The grandfather paradox requires that time-travel is impossible. The nihilist paradox requires that breaking an axiom-level loop of separators in the noumenal world is impossible — and this is confirmed by the fact that each separator in the loop has intrinsic necessity: they cannot not exist.

As a final note to our discussion of possible worlds and actual world, something should be said about **contingent possible worlds**, which are possible worlds in which at least one thing is contingent: contingent is the sense of being one of a plurality of possibilities. Suppose, for example, that somewhere in G some relation could be equally either $\hat{p}R$ or $\hat{q}R$; there would then be two possible G's, one with $\hat{p}R$ and the other with Q̂R, and since each would have a top relation having necessary actuality, each would be actual — which is impossible. So there cannot be any contingencies in G, G has to be fully determined. But possible worlds other than G may be contingent, even wildly so, and every contingency would require a plurality of possible worlds. Since these cannot be actual, they have ideal existence only, and are of no more interest to scientists and philosophers than are mermaids, griffins, centaurs, and dragons.

The origin of the idea of contingent worlds is our original definition of the minim as the minimal property set of a relation, which contains the three primitive properties of adicity, possibility, and simplicity. This possibility is either plural or singular, and leads to loop possibility or to loop necessity. A world with the first is then a contingent world and a world with the second is the actual world; and there are a plurality of axiom levels leading to a plurality of contingent worlds, and a singularity of axiom levels leading to the necessary world.

So among all possible worlds there is one that is actual, and hence truly describes the noumenal world, and the rest are contingent.

We abbreviate 'best of all possible worlds' to B.P. and the proposition that the B.P. exists necessarily because it is the best to B.P.P.

Chapter 9. Formal Proof.

In this chapter a formal proof is offered, using relational logic, that the best of all possible noumenal worlds exists necessarily because it is the best. The advantage of this is that it is much easier to show validity or invalidity in a formal proof than it is in an argument in words. We assume all the foregoing concepts of relation, properties, terms, uppers, structures, wholes, possible worlds, etc.

Abbreviations used in the proofs.

Assump.: assumption, the start of a conditional proof or an indirect proof (233).

C.A.: chain argument: $[(ʔp⇒ʔq)∧(ʔq⇒ʔr)]⇒(ʔp⇒ʔr)$.

C.P.: conditional proof.

Conj.: conjunction: $(ʔp∧ʔq)⇒ʔ(p∧q)$.

Def.: definition.

DeM.: DeMorgan: $ʔ(ʔp∧ʔq)⇔~(~p∨~q)$, $ʔ(ʔp∨ʔq)⇔~(~p∧~q)$.

D.N.: double negation: $~~p⇔ʔp$.

D.S.: disjunctive syllogism: $[(p∨q)∧~q]⇒p$, $[(p∨q)∧~p]⇒q$.

Equiv.: equivalence, $⇔$.

I.P.: indirect proof, also known as *reductio ad absurdum*, or proof by contradiction: $[p⇒(q∧~q)]⇒~p$

M.P.: *modus ponens*, or affirmation of the antecedent: $[ʔ(p⇒q)∧ʔp]⇒ʔq$.

Rep.: repeat.

Simp.: simplification: $ʔ(p∧q)⇒ʔp$, $ʔ(p∧q)⇒ʔq$.

Taut.: tautology.

Th.: theorem.

U.I.: universal instantiation.

U.G.: universal generalisation.

Fonts:

Times New Roman

U.C: relations, as in R, S.

U.C. italic: sets, as in *P*, *Q*.

L.C: propositions, statements, as in p, q, r.

Small caps: terms of relations, as in R, S.

Small caps, italic: sets of terms of relations, as in *R*, *S*.

Small caps, capped: properties of relations, as in R̂, Ŝ.

Small caps, capped, italic: sets of properties of relations, as in *R̂*, *Ŝ*.

Subscripts: denotation of particular individuals and scopes. The scopes, which apply to quantifiers and to uniqueness, are R, W, U, and P̂; R standing for all relations in all possible worlds, W standing for all of a possible world, U standing for all possible worlds and P̂ standing for all possible properties of a relation.

Reserved letters:

C: a consciousness.

M̂ : the minim: the three properties of adicity, possibility, and simplicity.

T: the top relation of a whole or of a possible world.

W: a whole.

W: a possible world.

Arial:

Reserved letters:

G: the best of all possible worlds.

T: the top relation of G.

U: unique. Subscripts denote the range, as above under "Subscripts".

Definitions used in the proofs.

Actuality: (140) intrinsic necessity, symbolised $\boxdot$.

Actual world: (140) the possible world that has loop actuality, $\boxdot\mathbf{W}$. It is complete because it is a candidate for being a true description of the noumenal world, which is complete by definition. In symbols this latter is $(\mathcal{E}\mathbf{W})[(\boxdot\mathbf{W}_1 \wedge \boxdot\mathbf{W}_2) \Rightarrow (\mathbf{W}_1 = \mathbf{W}_2)]$.

G: (44) the best of all possible worlds; best since having the highest #T.

T: (31) the top relation of G; hence GT.

$\mathsf{U_G}x$: x is unique in G; at least one x and at most one x are in G.

General notes.

We extend the notation for relations to that of possible worlds, where a property or a term apply to the whole world: for example, $\boxdot\mathbf{W}$ means that every relation in $\mathbf{W}$ has $\boxdot$ and $\mathbf{W}\mathsf{T}$ means that $\mathbf{W}$ has a top relation T.

A letter that has two different subscripts is assumed to refer to two different entities, as in $\mathbf{W}_1 \neq \mathbf{W}_2$, $R_1 \neq R_2$, $\hat{R}_1 \approx \hat{R}_2$.

A proof is followed, where needed, by line-numbered notes in a smaller typeface. Abbreviations are explained in such notes after their first occurrence, as well as on page 144. For the argument forms in the proofs, see Chapter 3, and for Conditional Proof and Indirect Proof, see Appendix B.

Existential quantification is written as $\Diamond\Phi$ rather than $(\mathcal{A}x)\Phi x$, where Φ refers to any quantifiable formula.

Uniqueness is symbolised by U and a subscript that indicates its range: thus $\mathsf{U_W}x$ means that x is unique among all possible worlds and $\mathsf{U_R}x$ means that x is unique among all possible relations. Uniqueness is proved by proving "at least one, and at most one".

Axioms.

1. $(\mathcal{E}R)(\mathcal{E}\mathbf{W})[(\boxdot R_R < \mathbf{W}) ⇕ \boxdot \mathbf{W}]$: Actuality in an instance of a relation R_R loop-necessitates the actual existence of the possible world that contains that relation R_R.

> This is plausible because an instance of a relation can be actual if and only if its terms are actual, with the exception of T, which has intrinsic necessary actuality $\boxdot\mathsf{T}$ (136).

2. $(\mathcal{E}\hat{P})(\mathcal{E}R)[\lozenge\hat{P}⇒(\lozenge R_R \wedge \hat{P}R_R)⇒(\lozenge\mathbf{W}>\hat{P}R_R)]$: Every possible property exists in at least one kind of a relation, and thereby in at least one possible world.

> This is plausible because if a property is possible then it exists, and it can only exist in a relation, so the relation exists. And every possible relation exists in at least one possible world because a possible world can be built up around it, by emergence, demergence and a suitable axiom level.

3. $(\mathcal{E}\mathbf{W})(\mathcal{E}\hat{R}R)\{[(\hat{R}R_A < \mathbf{W}_1\mathsf{T}_1)]⇒$
$(\lozenge\mathbf{W})(\lozenge\hat{R}R)[\mathbf{W}_2\mathsf{T}_2 \wedge (\#\mathsf{T}_2 \leqslant \#\hat{R}R) \wedge (\hat{R}R_B < \mathbf{W}_2\mathsf{T}_2)]\}$
If a relation $\hat{R}R_A$ having $\#\hat{R}R_A$ is possible in one possible world $\mathbf{W}_1\mathsf{T}_1$ then a similar relation, $\hat{R}R_B$, having $\#\hat{R}R_B=\#\hat{R}R_A$ is possible in another possible world, $\mathbf{W}_2\mathsf{T}_2$, provided that $\#\mathsf{T}_2 \geqslant \#\hat{R}R_A$.

> This is plausible because if a relation has an emergent level *n* in one possible world then a similar relation may emerge at that level in another possible world having that level.

Theorem 9.1. At least one possible world is actual: $\boxdot\mathbf{W}$.
Proof:

1. $\boxdot\mathbf{C}$		Premise.
2. $\boxdot\mathbf{C}\Rightarrow\boxdot R\textsc{r}$		Premise.
3. $\boxdot R\textsc{r}$		1, 2, M.P.
4. $\boxdot R\textsc{r}\Rightarrow\Diamond\boxdot$		Taut.
5. $\Diamond\boxdot$		4, 3, M.P.
6. $(\mathcal{E}\hat{P})(\mathcal{E}R\textsc{r})[\Diamond\hat{P}\Rightarrow(\Diamond R\textsc{r}\wedge\hat{P}R\textsc{r})\Rightarrow(\Diamond\mathbf{W}>\hat{P}R\textsc{r})]$ Axiom 2.
7. $\Diamond\boxdot\Rightarrow(\Diamond R\textsc{r} \wedge\boxdot R\textsc{r})\Rightarrow(\Diamond\mathbf{W}_1>\boxdot R\textsc{r})$ 6, U.I.
8. $\Diamond\boxdot\Rightarrow(\Diamond\mathbf{W}_1>\boxdot R\textsc{r})$		7, C.A.
9. $\Diamond\mathbf{W}_1>\boxdot R\textsc{r}$		8, 5, M.P.
10. $\boxdot R\textsc{r} <\Diamond\mathbf{W}_1$		9, Taut.
11. $(\mathcal{E}R)(\boxdot R<\mathbf{W}_1)\Updownarrow\boxdot\mathbf{W}$		Axiom 1.
12. $(\boxdot R\textsc{r}<\mathbf{W})\Updownarrow\boxdot\mathbf{W}_1$		11, U.I.
13. $(\boxdot R\textsc{r}<\mathbf{W})\Rightarrow\boxdot\mathbf{W}_1$		12, Taut.
14. $\boxdot\mathbf{W}_1$		10,13, M.P.

9.1.1   This premise is an empirical fact, established by Descartes' *cogito*: I am conscious therefore I exist. Thus a consciousness, **C**, indubitably exists — that is, it is necessarily empirically actual. **C** is the consciousness of whomever is considering this matter.

9.1.2   This premise is also an empirical fact: the consciousness **C** is necessarily consciousness *of* something, and this *of* is a relation, R, and this *something* is a term, $\textsc{r}$, of that relation; so $R\textsc{r}$ is actual.

9.1.3   M.P. stands for *Modus Ponens*, or affirmation of the antecedent: $[(p\wedge q)\wedge p]\Rightarrow q$.

9.1.4   Taut. stands for tautology: any tautology, being obvious, may be treated as an equivalence; obviously $\Diamond\boxdot R\textsc{r}$ is equivalent to $\Diamond\boxdot$. $\Diamond\boxdot$ means that actual existence is possible.

9.1.6   '>' here means 'has the part'.

9.1.7   U.I. stands for Universal Instantiation.

9.1.8   C.A. stands for Chain Argument, or Hypothetical Syllogism: $(p\Rightarrow q\Rightarrow r)\Rightarrow(p\Rightarrow r)$.

Theorem 9.2. At least one instance of a relation is actual:
$\Diamond(\boxdot R_R)$.
Proof:

1. $\boxdot R_R$	Th. 9.1.3.
2. $\boxdot R_R \Rightarrow \Diamond(\boxdot R_R)$	1, Taut.
3. $\Diamond(\boxdot R_R)$	2, 1, M.P.

9.2.1 A line from a previous proof may be entered as a line in a later proof, provided that it is not a restricted line, indented, as in Conditional Proof or Indirect Proof.

Theorem 9.3. At most one possible world has an instance of a relation that has actual existence:
$$(\mathcal{E}R)\{[(\boxdot R_1 < \mathbf{W}_1) \wedge (\boxdot R_2 < \mathbf{W}_2)] \Rightarrow (\mathbf{W}_1 = \mathbf{W}_2)\}.$$
Proof:

1. $(\mathcal{E}\hat{P})(\mathcal{E}R_R)[\Diamond \hat{P} \Rightarrow (\Diamond R_R \wedge \hat{P} R_R) \Rightarrow (\Diamond \mathbf{W} > \hat{P} R_R)]$	Axiom 2.
2. $\Diamond \boxdot \Rightarrow (\Diamond R_1 \wedge \boxdot R_1) \Rightarrow (\Diamond \mathbf{W}_1 > \boxdot R_1)$	1, U.I.
3. $\Diamond \boxdot$	Th. 9.1.5.
4. $(\Diamond R_1 \wedge \boxdot R_1) \Rightarrow (\Diamond \mathbf{W}_1 > \boxdot R_1)$	2, 3, M.P.
5. $\boxdot R_1$	Assump. C.P.
6. $\boxdot R_1 \Rightarrow \Diamond R_1$	Def. of actuality.
7. $\Diamond R_1$	6, 5, M.P.
8. $\Diamond R_1 \wedge \boxdot R_1$	7, 5, Conj.
9. $\Diamond \mathbf{W}_1 > \boxdot R_1$	4, 8, M.P.
10. $\boxdot R_1 < \Diamond \mathbf{W}_1$	9, Equiv.
11. $(\boxdot R_1 < \Diamond \mathbf{W}_1) \Downarrow \Uparrow \boxdot \mathbf{W}_1$	Axiom 1, and U.I.
12. $\boxdot \mathbf{W}_1$	11, 10, M.P.
13. $\boxdot R_1 \Rightarrow \boxdot \mathbf{W}_1$	5-12, C.P.
14. $\boxdot R_2 \Rightarrow \boxdot \mathbf{W}_2$	Rep. 2-13, $R_2 \mathbf{W}_2$.
15. $\boxdot R_1 \wedge \boxdot R_2$	Assump. C.P.
16. $\boxdot R_1$	15, Simp.
17. $\boxdot \mathbf{W}_1$	13, 16, M.P.
18. $\boxdot R_2$	15, Simp.
19. $\boxdot \mathbf{W}_2$	14, 18, M.P.
20. $\boxdot \mathbf{W}_1 \wedge \boxdot \mathbf{W}_2$	17, 19, Conj.
21. $(\mathcal{E}\mathbf{W})[(\boxdot \mathbf{W}_1 \wedge \boxdot \mathbf{W}_2) \Rightarrow (\mathbf{W}_1 = \mathbf{W}_2)]$	Def. of $\boxdot \mathbf{W}$.

22. $(\boxdot W_1 \wedge \boxdot W_2)] \Rightarrow (W_1 = W_2)$ 21, U.I.

23. $W_1 = W_2$ 22, 20, M.P.

24. $(\boxdot R_1 \wedge \boxdot R_2) \Rightarrow (W_1 = W_2)$ 15-23, C.P.

25. $(\boxdot R_1 < W_1)$ Th. 9.1.10.

26. $(\boxdot R_2 < W_2)$ Th. 9.1.10.

27. $(\boxdot R_1 < W_1) \wedge (\boxdot R_2 < W_2)$ 25, 26, Conj.

28. $[(\boxdot R_1 < W_1) \wedge (\boxdot R_2 < W_2)] \Rightarrow (W_1 = W_2)$ 24, 27, Taut.

29. $(\mathcal{E}R)\{[(\boxdot R_1 < W_1) \wedge (\boxdot R_2 < W_2)] \Rightarrow (W_1 = W_2)\}$ 24, U.G.

9.3.3   Intrinsic actuality is possible, $\Diamond \boxdot$, because $\boxdot W$.

9.3.5   Indentation of lines, such as lines 5-12, means that they are dependent on an assumption, (C.P. or I.P.) and may not be referred once that proof is completed.

9.3.6   Necessity is defined as singular possibility, which is a special case of possibility.

9.3.8   Conj. stands for Conjunction.

9.3.10   Whatever X and P may be, 'X has a part P' is equivalent to 'P is a part of X'.

9.3.14   "Rep. 2-13, R_2, W_2," means repeat lines 2 to 13, using R_2, W_2.

9.2.16   Simp. stands for Simplification: $(p \wedge q) \Rightarrow p$.

Theorem 9.4. An instance of a relation having actual existence is unique to one possible world. $U_w \boxdot R_R$.
Proof:

1. $\Diamond (\boxdot R_R)$ Th. 9.2.

2. $(\mathcal{E}R)\{[(\boxdot R_1 < W_1) \wedge (\boxdot R_2 < W_2)]q(W_1 = W_2)\}$ Th. 9.3.

3. $[(\boxdot R_1 < W_1) \wedge (\boxdot R_2 < W_2)] \Rightarrow (W_1 = W_2)$ 2, U.I.

4. $U_w \boxdot R$ 1, 3, Def. of unique.

Theorem 9.5. T, the top relation of the best of all possible worlds, has intrinsic actuality: $\boxdot T$.
Proof:

1. $(\boxdot R_A \neq T) \wedge (\boxdot R_A < W_1 T_1)$ Assump. I.P.

2. $(\mathcal{E}W)(\mathcal{E}\hat{R}R)\{[(\hat{R}R_A < W_1 T_1)] \Rightarrow$
   $(\Diamond W)(\Diamond \hat{R}R)[W_2 T_2 \wedge (\#T_2 \leqslant \#\hat{R}R_A) \wedge (\hat{R}R_B < W_2 T_2)]\}$
        Axiom 3.

3. $(\boxdot R_A < W_1 T_1) \Rightarrow$

$[\mathbf{W_2T_2} \wedge (\#\mathbf{T_2} \leqslant \#\hat{R}RA) \wedge (\square RB < \mathbf{W_2T_2})]$ 2, Inst.

4. $\square RA < \mathbf{W_1T_1}$ ____ 1, Simp.

5. $\mathbf{W_2T_2} \wedge (\#\mathbf{T_2} \leqslant \#\hat{R}RA) \wedge (\square RB < \mathbf{W_2T_2})$ 3, 4, M.P.

6. $(\square RB < \mathbf{W_2T_2})$ ____ 5, Simp.

7. $(\square RA < \mathbf{W_1T_1}) \wedge (\square RB < \mathbf{W_2T_2})$ ____ 4, 6, Conj.

8. $(\square RA < \mathbf{W_1T_1}) \wedge (\square RB < \mathbf{W_2T_2}) \Rightarrow \sim U_w \square R$
____ 4, Def. of unique.

9. $\sim U_w \square R$ ____ 8, 7. M.P.

10. $U_w \square R$ ____ Th. 9.4.

11. $U_w \square R \wedge \sim U_w \square R$ ____ 9, 10, Conj.

12. $\sim[(\square RA \neq T) \wedge (\square RA < \mathbf{W_1T_1})]$ ____ 1-11, I.P.

13. $\sim(\square RA \neq T) \vee \sim(\square RA < \mathbf{W_1T_1})$ ____ 12, DeM.

14. $(\square RA = T) \vee \sim(\square RA < \mathbf{W_1T_1})$ ____ 13, D.N.

15. $\square \mathbf{W}$ ____ Th. 9.1.

16. $\square \mathbf{W} \Rightarrow \square \mathbf{W_1T_1}$ ____ 15, Taut.

17. $\square \mathbf{W_1T_1}$ ____ 16, 15, M.P.

18. $\mathbf{W_1T_1} > \square RA$ ____ Th. 9.1.9.

19. $\square RA < \mathbf{W_1T_1}$ ____ 18, Taut.

20. $\sim\sim(\square RA < \mathbf{W_1T_1})$ ____ 19, D.N.

21. $\square RA = T$ ____ 14, 20, D.S.

22. $\square RA$ ____ Th. 9.1.3.

23. $\square T$ ____ 21, Equiv.

9.5.1 Assump. I.P. stands for Assumption Indirect Proof. See Appendix B. '$\neq$' means 'not identical'. Note that lines 1-11 are indented to show that they may not be quoted outside this indirect proof.

9.5.3. "Inst." here stands for E.I., twice, and U.I., twice; E.I. stands for existential instantiation.

9.5.12 1-11, I.P. stands for Indirect Proof, using lines 1-11.

9.5.13 DeM stands for DeMorgan.

9.5.14 D.N. stands for Double Negation; in this case $\sim(n \neq m) \Leftrightarrow (n=m)$.

9.5.21 D.S. stands for Disjunctive Syllogism.

Theorem 9.6. The best of all possible worlds is actual: $\boxdot G$.

Proof:

1. $(\mathcal{E}R)(\mathcal{E}W)[(\boxdot R{<}\mathbf{W_1})⇪\boxdot\mathbf{W_1}]$:	Axiom 1.
2. $(\boxdot T{<}G)⇪\boxdot G$	1, U.I., twice.
3. $\boxdot T$	Th. 9.5.
4. $T{<}G$	Defs. of T and G.
5. $\boxdot T{<}G$	3, 4, Equiv.
6. $\boxdot G$	2, 5, M.P.

9.6.6   ⇪, being necessitous, may be treated as $\Rightarrow$.

Chapter 10. B.P.P. and Physics.

       B.P.P., the principle that the best of all possible worlds exists necessarily because it is the best, appears to be significant for physics, and for science in general, in a number of ways, but I am neither a scientist nor a mathematician, and may well be rushing in where angels fear to tread; so I offer this chapter with deference to the experts.

       The need to figure out the number of kinds of actual noumenal separators, and their properties, and hence an absolutely fundamental theory of physics, is, needless to say, a wonderful opportunity for fame and fortune for some genius. What is offered here is a very pale shadow of what is needed, but it nonetheless serves to illustrate how the concept of separator can lead, *via* cascading emergence, to an applied mathematics, to B.P.P., and to the Grand Structure.

       Twelve points are considered.

1.   The quantum-mechanical experimental violation of Bell's (1928-1990) inequality theorem can be accounted for by denying one or more of three assumptions, called by physicists locality, realism, and logic. Locality, as Einstein put it, is that there is no spooky action at a distance; realism is the claim that there is existence without observation — which in the present context means that there is noumenal existence (28); and logic includes mathematics. To deny any one of these three is very difficult, or else impossible; but so is Bell's inequality. However, as Bell himself pointed out,

> "... one of the ways of understanding this business is to say that the world is super-deterministic. That not only is inanimate nature deterministic, but we, the experimenters who imagine we can choose to do one experiment rather than another, are also determined. If

153

so, the difficulty which this experimental result creates disappears."[15]

Thus there is another possible solution to the paradox: strict determinism. This, for Bell, is more than material determinism, it is determinism of *everything*, which means that there is no human freedom, no freedom of the will. Thus our sensation of being free, of making decisions freely, is illusory. (In philosophy, the word determinism means this strict determinism.) If the noumenal world has no action at a distance, and is real, logical, and deterministic, then the paradox of Bell's inequality vanishes. But this is just what B.P.P. requires: the real world — the noumenal world — has causality and loop necessity, exists independently of perception and is mathematical — that is, relational. And, also, everything in this best of all possible worlds is determined: if it were not then there would be at least one undetermined something in B.P. — which is impossible (142). Thus there is nothing random in the best of all possibles. The moment when a particular radioactive atom fissions may be contingent in the sense that it is uncaused, but it is still determined in that it is downward or upward necessitated: causation is horizontal necessitation, as opposed to upward or downward necessitation.

Another way of looking at this is that the statistical interpretation of the wave function, Ψ, as being a probability, $|\Psi|^2$, originated by Max Born (1882–1970), is incorrect. A mathematical probability is not necessarily a statistical probability: we have an example of this in the definition of hekergy as $\ln.(t/e)$ and e/t is a probability but not a relative frequency. (Notice that since hekergy, H, is $\ln.(t/e)$, it follows that $t/e = e^H$ so $e/t = e^{-H}$; e/t is a probability so if we put $e/t =$

---

[15] P. C. W. Davies and J. R. Brown, *The Ghost in the Atom: A Discussion of the Mysteries of Quantum Physics*, Cambridge University Press, 1986/1993, p. 47.

$|\Psi|^2$ then $|\Psi|^2 = e^{-H}$, hence $|\Psi| = e^{-H/2} = e^{i(H/2)}$. This is highly suggestive, but I must leave the development of it to physicists. Note that H here is a hekergy, not the Hamiltonian.)

This determinism explains the solution to the old philosophical problem of free will, which is: a free action or decision is an event which, like all events, must either be necessitated or else not be necessitated. If it is necessitated then it is not a free action, and if it is not necessitated then it is a chance event and so not willed, hence not a free action. Therefore there is no free will. Our sensation of being free must consequently be an illusion — probably best explained as a product of vanity. Another way of looking at this is that the present moment is a boundary between a changeable future and an unchangeable past; and passage of time is the movement of this boundary into the future. But how could a boundary be an operator that converts changeable to unchangeable? It cannot: hence the future must be unchangeable, therefore determined. Those who find all this intolerable will no doubt believe otherwise; but believing known falsehoods, although irrational, is not a crime in civilised countries.

2.   The collapse of the wave-function is downward necessitated: a necessity collapses the wave-function because on a higher level an observer, or a measuring device, or something unknown, requires it.

3.   Wave-particle duality can be explained in terms of B.P.P., which requires that the noumenal world consists entirely of relations. So if a wave-particle is a relation, P, in the noumenal world then its terms must be relations, and these terms must have various configurations which either cause the particle to emerge (emergence-configurations), or not (submergence-configurations). These configurations of terms may be supposed to be constantly changing but the changes cannot be random — there is nothing random in B.P. — so

they must be repetitive in some pattern and so form waves: standing waves. Thus we can think of the configurations of the terms of the relation P being described by the wave-function. However the wave function may also describe submergence-configurations of the terms of the particle so that the particle submerges. When this happens the particle P has to submerge — cease to exist — but because it has energy it can only do so for a short time, within the limits of the uncertainty principle, analogous to the existence of a virtual particle, except that here the relation becomes a virtual non-particle. If the energy of P is E then its submergence duration, Δt, is such that $E\Delta t \leq \hbar$. Thus in situations in which the particle/relation P cannot go, such as through a double slit, tunnelling through a potential barrier, or making a quantum leap from one energy level to another in an atom, the particle may submerge for long enough for its terms to go through and then have the particle re-emerge in the new location; it is a virtual non-particle for the transition. And the terms go through as a wave.

A concrete example will clarify this. Suppose that my hat is *on* the table, and that there are four positions on the table that the hat may have: left, back, right, and front. Each position is a different emergence-configuration of the hat and the table, the hat and the table being the terms of the relation *on*. Suppose also that my hat may be *on* my head, in one of four emergence-configurations: tilted to the left, back, right, and front. This second relation *on* is the same *kind* of relation as the first but a different *instance*. They are the same kind because, like electrons, they are indistinguishable, they have similar property sets; but they are different instances because they have different terms — {hat, table} and {hat, head}. However, in taking my hat from the table in order to put it on my head the first *on* submerges and then later a new *on* emerges, having the terms {hat, head}. The relation *on* did not exist during the transition. If now it is necessary that the emergence-configurations of {hat, table} and of {hat, head} have to keep changing, cyclically through left, back, right, and front then this pattern of changing emergence-configurations is a standing square wave. Increase the number of members of both of these sets of emergence-configurations and the waves increasingly approach sinusoidal waves. Thus the relation *on* is both a particle and a wave. Furthermore, if I twirl around as I dance a Viennese waltz, while wearing my hat, there will be other waves, and super-positioning, for the relation *on*; and if, as I circle around the dance floor, these lesser waves are constrained by the requirement that my path around the floor has a length that is an integral number of half wavelengths, then my possible paths will be quantised. So as I changed from one possible path to another the hat would have to fly of my head as I left the first path and land back on my head as I reached the other path; and because of this the relation *on* would not exist during this quantum jump, although hat and head — the terms of the *on* — would continue to exist.

  We saw (101) that given atomic length separators and temporal separators which are somewhat elastic, gravitational travelling waves would be possible. So if these separators — or, better, atomic space-time separators — have more, suitable, properties than those required by general relativity, then they could emerge electromagnetic and other travelling waves. Alternatively, these waves could emerge at higher levels, in which the necessary mathematical properties were also emergent. Putting this another way, it might be that space-time separators in the axiom level emerge a gravitational field in level-2; emergent space-time separators in level-2 — atomic squares — form an electromagnetic field in level-3; and emergent space-time separators in level-3 — atomic cubes — form a strong force field in level-4. Thus variety increases with level.

4. String theory is possible in B.P. Imagine a piece of paper covered in plus signs, in regular order, horizontally and vertically. These are linear separators in a two dimensional space. The horizontal separators must form closed loops,

because otherwise the ones at the vertical edges of the paper would each lack one term and so could not exist, which would submerge the ones next to them, and so on across the paper. So the paper would have to be curved into a tube. And the vertical separators also must form closed loops, which would require the tube to be curved into a torus. The minor circumference of the torus could then constitute a new dimension. The linear separators in this torus would have to be distorted in length, in order to form the torus, so would necessarily be elastic — in which case, given temporal separators, they could both vibrate and provide a curvature to space-time.

The elasticity of the separators need not be large: if, say, it allowed a stress of one part in 10^{20} and a linear separator was a Planck length (242) then there would have to be 10^{20} Planck lengths to allow a compression of one such length. And, as we know, there is a large number of Planck lengths in the Universe.

5.　　　　B.P.P. is significant in physics also in that it denies infinities, including infinite divisibility. This means that mathematical physics should stop using the real number system in favour of the rational number system, and deny the decimal infinities of rational numbers. In other words, no measurable quantities are infinitely divisible and no measurable quantities are infinite in extent. So every measurable quantity has a minimum magnitude and a maximum magnitude. We see some of this immediately if separators have Planck units as minimum magnitudes: an atomic length is a Planck length, l_p, and an atomic duration is a Planck time, t_p. (See Appendix D for more on the Planck units.) And the maximum velocity is one Planck length per Planck time, which is c, the velocity of electromagnetic radiation. Other maxima and minima may be calculated or inferred from other Planck units. For example, a minimum change — a dissimilarity over time — occurs over one t_p, so a minimum cycle requires $2t_p$ — such a change, and back again — hence the maximum possible frequency is $v_p = 1/2t_p$, which is $10^{44}/(2\text{x}5.391)$, or v_p is $9.275 \text{x} 10^{42}$. (All these amounts are approximate.) Equally, $\lambda v = c$ for a photon so if the minimum wavelength $\lambda_p = 2l_p$ then v_p is $c/2l_p = 9.273 \text{ x} 10^{42}$ approximately. Because the energy of a photon is $E = \hbar v$ and $\hbar = 1.055\text{x}10^{-34}$ J.s it would follow that the maximum energy of a photon is $1.054 \text{x} 10^{-34} \text{x } 9.275 \text{x} 10^{42}$ or $9.775\text{x}10^{10}$, which is approximately 10^{11} Joules. But the Planck energy, E_p, is $1.9561\text{x}10^9$ J, or approximately $2 \text{x} 10^9$ J. So the maximum energy of a single photon is $50 \ E_p$.

Separators are such that they must form closed loops (31). So the total series of temporal separators form a (finite) closed loop, and hence the Big Bang and the Big Crunch are the one identical event, sometimes called the Big Bounce. The Universe does not cycle from one big bang to another, to another, repeatedly, like a helix; it just circles from the big bang to the big bang, and this one circle is the totality of time. More generally, space-time must form a loop, so has to be a four-dimensional sphere.

B.P.P. provides the major units — mass, length, and time — for dimensional analysis (see Appendix D). And it also suggests that causal set theory and loop quantum gravity have the best chances of solving the incompatibility between general relativity and the standard model of quantum mechanics, since these former two theories both assume a quantisation of space-time.

Also, given the denial of infinities, there are no mathematical singularities in the best possible noumenal world so the Big Bang did not start as a singularity, a black hole is not a singularity and every particle has a size.

Finally, infinities are a bother with re-normalisation, and quantum gravity; so their denial may cure those bothers.

6.   That the noumenal world is the best possible requires that it has maximum possible hekergy. This in turn requires

certain things in physics. The hekergy of a relation was defined as ln.(t/e), where e is the number of possible emergence-configurations of the relation and t is the total number of its possible configurations and $t>1$. So if $e=1$ then the hekergy of the relation is a maximum. We have already considered stationary principles in physics, all of which require a relation to have $e=1$.

Maximum hekergy also means maximum value in human terms: values such as the traditional truth, beauty, and goodness. Truth and beauty are both criteria of good explanation (Chapter 7), so maximum hekergy in B.P. means that a theory that has maximum beauty must have maximum truth, and *vice versa*. This is why we find beauty in nature: not only beauty in such things as birds, butterflies, tigers, and sunsets, but also in symmetries. And maximum goodness means that in principle there is a scientific basis for ethics: all that is needed for a start in this is a way of measuring goodness, which is a form of hekergy.

7.   B.P.P. also offers the best solution to the problem of cosmic coincidences, also called the fine turning problem. This is the problem that there are a number of parameters in theoretical physics which are only known empirically: there is no theoretical basis for them having the values that they do. Among these are the rest-masses of the various wave-particles and the ratios of the strengths of the four fundamental forces. All of these parameters could have different values, according to theory, but if they did then life in our Universe would be impossible because the lives of stars would be too short, or galaxies could not form, or supernovas making heavier elements could not form, etc. Explaining the fact that these parameters all do have values that allow life is the problem of cosmic coincidences. It is a problem because current theories do not satisfactorily explain the coincidences, and they do need to be explained. Each of these actual values is in itself

quite improbable, and the totality of them together is vastly more so.

The problem is frequently expressed in terms of the Anthropic Principle: because it is an empirical fact that we exist, our universe must have parameters having these particular values; the question is: why does it have them? As we have seen, explanation of empirical facts is by means of theoretical noumenal causes of them, so what are the causes of all these parameters having these particular values? The Anthropic Principle seems at first sight to be quite trivial, but it is a little more than trivial if it is understood as: the fact of our existence demerges the favourable parameters. In other words, the Grand Structure requires the cosmic coincidences.

There are basically two current explanations of cosmic coincidences, both unsatisfactory. One is the many-world hypothesis and the other is the fine-tuner hypothesis.

The many-world hypothesis is that in the one Universe (noumenal world) there are a very large number of mutually isolated mini-universes, each taking on its own values of the parameters, at random. Given a huge number of these mini-universes, at least one will have the right values of its parameters so that stars, galaxies, and life are possible. Our own mini-universe has stars, galaxies, and life, and so is one of these very improbable ones. The two main objections to this are that it multiplies mini-universes far beyond necessity, contrary to Occam's Razor, and, second, that there is no way in which the existence of these mini-universes could be either verified or falsified, so that the postulation of them is quite unscientific. Also, from the point of view of B.P.P. there is nothing random. And we have seen that there cannot be more than one noumenal world, so these mini-universes could only be structures, unrelated to each other, in one possible world; but this would not be the best possible world, so is here denied.

Part of the appeal of the many-world hypothesis may be its close resemblance to the epistemological requirement of many possible ideal noumenal worlds but only one actual noumenal world. But to suppose from this that everything possible is actual is an error: it is impossible for more than one possible world to be actual.

The fine-tuner hypothesis is that these cosmic coincidences are a fine-tuning of the Universe, metaphorically like the tuning of a piano or a violin, and fine-tuning implies a Tuner, or God, who tuned or willed the world to be as it is. The objection to this explanation is that, to paraphrase Pierre-Simon Laplace (1749-1827), physics has no need of the God hypothesis — a theory that has very low density of detail, does not harmonise with physical theories, and does not predict empirical novelties. In short, it is neither falsifiable nor verifiable (121). Even worse, as an explanation it merely pushes the existence problem back a stage, by raising the question as to why there should be a Tuner.

A much better explanation appears first in the work of Gottfried Wilhelm Leibniz (1646-1716), who said that the actual world exists necessarily because it is the best of all possible worlds (214). He was widely misunderstood on this because everyone assumed that he was referring to the world of empirical phenomena that we all experience around us. But he was not: he was referring to the noumenal world, as opposed to the empirical world of everyday experience. Obviously the best of all possible worlds would be less than the best if it lacked stars, galaxies, and life. The B.P.P. explanation does not require an innumerable number of unobservable mini-universes, nor a *deus ex machina*, and so wins out over the other two.

8.   Quantum entanglement must be a relation — if you think about it, could entanglement be anything else but a relation? If two particles are entangled then, being two, there is some distance between them — between the terms of the entanglement relation — but there need not be any restriction on the finite magnitude of this distance. A relation must

extend to all of its terms, as a simple example shows: if a married couple live in New York and one of them travels to Los Angeles then they are separated by a continent, and the relation of wedlock that holds between them extends across that same continent. The spatio-temporal magnitude of a relation is the separation of its terms.

9.   Pervasion is another point clarified by B.P.P. In the days of classical physics it was supposed that there was a luminiferous ether, calculated to be a medium of enormous density and elasticity, which *pervaded* all of space and time, and which explained the wave nature of electromagnetic radiation — since there cannot be a wave without an elastic medium in which it travels. That idea is now stone dead, but pervasion continues, in that fields such as the gravitational field, the electromagnetic field, and the Higgs field are supposed to *pervade* all of space-time. But pervasion does not make sense. If it exists in the noumenal world then it can only be a relation, in which case what are its terms? They can only be all of space-time, and all of a field, and this is multiplying entities beyond necessity, to say nothing of the fact that such a relation is not a medium. No medium pervades all of space-time, other than space-time itself. This is shown by the gravitational field, which is a distortion of space-time, not something that pervades space-time. So the electromagnetic and Higgs fields must also be distortions of elastic space-time, perhaps at a higher level than the gravitational, as with the level-2 structures called atomic squares and level-3 atomic cubes (100).

10.   B.P. also throws light on mathematics and logic in science. First, there is the question about mathematics: why does it apply to the world? A famous paper by Eugene Wigner (1902-1995), "The Unreasonable Effectiveness of

Mathematics in the Natural Sciences"[16] expresses his
puzzlement with this. The paper ends with:

> "The miracle of the appropriateness of the
> language of mathematics for the formulation of
> the laws of physics is a wonderful gift which
> we neither understand nor deserve. We should
> be grateful for it and hope that it will remain
> valid in future research and that it will extend,
> for better or for worse, to our pleasure, even
> though perhaps also to our bafflement, to wide
> branches of learning."

And Ian Stewart (1945-), in his "Why Beauty is Truth"[17]
writes:

> "The relationship between mathematics and physics is
> deep, subtle, and puzzling. It is a philosophical
> conundrum of the highest order — how science has
> uncovered apparent "laws" in nature, and why nature
> seems to speak in the language of mathematics."

Since the noumenal world consists entirely of relations,
and mathematics is our language of relations, and theoretical
science seeks to explain the empirical world by describing its
noumenal causes, it follows that theoretical science is
necessarily mathematical. The puzzle for Wigner and Stewart
— and for most scientists — arises from identification of
empirical world and noumenal world — an identification that

---

[16] Wigner, Eugene: "The Unreasonable Effectiveness of
Mathematics in the Natural Sciences," in *Communications in Pure and
Applied Mathematics*, vol. 13, No. I (February 1960). New York: John
Wiley & Sons, Inc.

[17] Ian Stewart, "Why Beauty is Truth: A History of Symmetry",
Basic Books, New York, 2009, p.275.

philosophers call **naive realism**. This is a case of the identity error (69): they cannot be one, they must be two, because the empirical world is perceptible and the noumenal world is not perceptible.

Second, and closely related to this topic, is the question of why it is that physicists have no use for formal logic. Whether it be the traditional logic of classes of Aristotle or modern set theory and quantificational logic — classes are sets, after all — it seems that neither has any explanatory value. This is no doubt the reason for the failure of past metaphysical systems — a failure that led to the scorn of the logical positivists — but we need to know why the logical approach failed while the mathematical one succeeded. Past metaphysicians used substance-attribute logic successfully to explain empirical things and qualities as noumenal substances and attributes, but invariably had trouble with empirical relations and hence numbers. Modern truth-functional logic explains nothing because it is a travesty of logic: it has the appearance but not the reality of genuine logic (64, 165). And set theory has explained numbers and relations for over a century, but in a somewhat unsatisfactory and cumbersome manner. Mathematics, on the other hand, has none of these problems. (We have not yet relationally explained concrete empirical qualities — secondary qualities, as they are called (175) — but will do in the next Chapter.)

There is also a minor point to be made concerning sets in science. Classification, even though it has no explanatory value, is useful in organising empirical data, as in biological, geological, and astronomical taxonomy. But it is not as useful at organising as is the organisation of numerical data, which leads to mathematical formulae.

11.   There are two major philosophic problems of science: the problem of induction in empirical science and the problem of verification in theoretical science, both of which are now soluble.

165

The **problem of induction**, which goes all the way back to Aristotle, is that to argue from some to all is logically invalid. The traditional example was that "All swans are white", and it demonstrated the fallibility of induction (107). The problem of induction in philosophy of empirical science — probably the most important problem in this field — is that empirical laws are arrived at inductively. We want to say that these laws are true, while other inductive generalisations, such as those of superstition, stereotypical thinking, and prejudice are false. But how can we justify such a distinction? In the past it was sometimes justified by the principle of uniformity of nature, but this principle was either itself arrived at inductively or else was a metaphysical fiat; and, either way, it also justified superstition and stereotypes. However B.P.P. justifies it, not by metaphysical fiat but by the wealth of necessity in the best of all possible worlds. There is no necessity in superstition, stereotypical thinking, and prejudice. The most widespread superstition in the western world is the belief that to speak of misfortune or evil brings it about — unless this is prevented by touching wood or crossing one's fingers.

The **problem of verification** applies to theories and is also a problem of logical fallacy. In its most general form it is $T \Rightarrow E$, or a particular theory, T, implies an empirical fact, E. If $T \Rightarrow E$ is analytically true (and this can be demonstrated logically) then there are two valid deductions that can be made from it. One is *modus ponens*, or affirmation of the antecedent, which is $[(T \Rightarrow E) \wedge T] \Rightarrow E$, or if both $T \Rightarrow E$ and T are true then E has to be true. The other is *modus tollens*, or denial of the consequent, which is $[(T \Rightarrow E) \wedge \sim E] \Rightarrow \sim T$, or if $T \Rightarrow E$ is true and E is false then T has to be false. A third argument form using $T \Rightarrow E$, frequently used but invalid, is called the fallacy of affirmation of the consequent; $[(T \Rightarrow E) \wedge E] \Rightarrow T$, or if both $T \Rightarrow E$

and E are true then T is true[18]. This last is the fallacious use of empirical data, E, to "verify" a theory, T; but then, how can a theory be verified? The solution to the problem of verification is that a theory is increasingly verified by increase in its quality of explanation, as discussed in Chapter 7; particularly so with successful theoretical predictions of empirical novelty.

12.　　　　Next, we consider principles in science. They are axioms, from which theories may be deduced. For some time the principles were usually principles of conservation: conservation of energy, of momentum, of electric charge, etc., but these are now derived from symmetries (which are relations described in group theory) via Noether's (1882-1935) theorem. There are other principles, called stationary principles, which are principles of maxima, minima, or points of inflexion in the rates of change of measurable quantities: for example, paths of least action, least time, or least resistance. Another principle is the principle of parsimony of hypothesis, which in philosophy is called Occam's Razor and was first proposed by William of Occam (1287–1347): do not multiply entities beyond necessity, meaning: do not invent more theoretical entities than are needed to explain the empirical facts. All of these principles can be derived from B.P.P., so B.P.P. might be called the first principle of science. It might be claimed to be the first principle of metaphysics, also; so something must be said about metaphysics.

　　　　If one looks at the history and nature of metaphysics sympathetically, instead of from a biassed logical-positivist point of view, then metaphysics is an attempt to know the

---

[18] The logical correctness of these three statements, plus a second fallacy, may be illustrated with any T⇒E that is obviously true, such as "If Pat is pregnant then Pat is female", from which it is clearly necessary that if a mammal named Pat is pregnant then she is necessarily female, while if Pat is not female then necessarily he is not pregnant; and if female she may or may not be pregnant, while if not pregnant Pat may or may not be female.

noumenal world: because metaphysicians want to explain the empirical world. But theoretical science is also an attempt to know the "underlying", or noumenal world, in order to explain the empirical world. So what is the difference between them? One difference is that theorists are specialists and metaphysicians are generalists: it is said that specialists are people who know more and more about less and less until they know everything about nothing; equally may it be said that generalists are people who know less and less about more and more until they know nothing about everything. More pertinently, past metaphysicians lacked the data of empirical science — measurements, particularly, and other experimental results — and had to rely on logic and the everyday empirical world to discipline their metaphysical speculations; theoretical scientists, on the other hand, have plenty of empirical data to discipline their theoretical speculations. There is a back and forth process of empirical data feeding theory and theory feeding ideas of new experiments. But the principle that is absolute core of both theoretical science and metaphysics is that *the noumenal world must be rational*. We can justify this.

Rationality is maximum hekergy in thought; it is so because it is necessitous; the necesssity is singular possibility; singular possibility means that $e=1$ in the hekergy of a proposition, which means that that hekergy is the maximum possible. So, since hekergy is value, rational thought is the most valuable kind of thought. And the noumenal world, being the best, is the most valuable of all possible worlds. So the noumenal world must be rational.

The noumenal world is rational, initially and ideally, in two ways: it does not contain any contradictions; and it does contain causes, which are noumenal analogues of logical necessities. However there are two kinds of rationality: mathematics and logic. The metaphysics of Plato was mathematical, and that of his student, Aristotle, was logical — the logic that Aristotle himself formulated — the logic of subjects and predicates, which led to a metaphysics of

168

substances and attributes. The empirical world consists of things, qualities, and relations. Things and qualities, which are concrete, are explicable by substance-attribute logic; and relations, which are abstract, are accounted for by mathematics. But Aristotle's logic cannot satisfactorily handle relations. Spinoza and Leibniz both solved this problem in different ways. Spinoza (213) supposed that there is only one substance, and all relations are among the attributes of this one substance. Leibniz (214) supposed that there are an infinity of substances, with no relations between them or within them. In each case they both finished up with maximum generality and minimum detail, so their solutions are unsatisfactory. Galileo, on the other hand, left all metaphysics to philosophers (or so he thought) and concentrated on the mathematics of Nature, the empirical world, which for him was a book to be read by scientists, written in the language of mathematics. But Galileo's most important contribution to the history of ideas was his principle of inertia, which Newton, *via* Descartes, made into his first law of motion, and this principle is both mathematical and a speculation about the noumenal. Galileo thought it to be an empirical law, and so did Newton; but it is a principle, and a better principle than Aristotle's principle that motion stops unless maintained by an external force. Consequently, with Galileo science became mathematical. Or, at least, more mathematical, since the science of astronomy had been mathematical from its beginnings.

So it is arguable that the most general of scientific theories and the least general of mathematical metaphysics must overlap. In other words, the difference between theoretical science and metaphysics is one of degree, not of kind. That said, it was suggested above that B.P.P. might be the first principle of both science and metaphysics, but there is a better one. The first principle is that there is a noumenal world and it is rational — mathematical — and real. We accept this principle because of our need to explain.

Traditionally, the main division of philosophers is into two classes: empiricists and rationalists. The significance of this for physics is that Niels Bohr was *par excellence* an empiricist and Einstein was *par excellence* a rationalist. An empiricist is one who believes that all knowledge comes through the senses, so that either there is no knowledge of that which cannot be perceived, or else that which cannot be perceived does not even exist. A rationalist is one who believes that that which exists but cannot be perceived can be known, using reason; what is so known is the subject of metaphysics, and how it is known is the subject of epistemology. Thus theoretical science is metaphysical.

Clearly, the present book has been rationalist from the very beginning, with its assumption of a noumenal world. So the Bohr-Einstein dispute, in the present context, should be decided in favour of Einstein.

However, one key point remains: what is the precise relationship between the empirical world and the noumenal world? The empirical-noumenal distinction is based on the perceptible-imperceptible distinction and hence on the nature of perception. So in order to understand the relationship between the noumenal and the empirical we need to enquire more into perception; this is the subject of the next chapter.

Chapter 11. Noumenal and Empirical.

We next need to examine the relationship between the two worlds, the empirical world of relations between concrete objects, which are structures of empirical qualities, and the underlying, abstract, noumenal world of imperceptible causes of these phenomena, and of other theoretical, hypothetical, entities — all of which are relations. The key concept relating these two worlds is explanation: as was remarked earlier, explanation is causal, causation is noumenal, the noumenal is imperceptible, we cannot manage without explanations, so we postulate the noumenal. And to deny all explanations is to fall into **solipsism**, the doctrine that I alone exist. (See Appendix C).

The relationship between the noumenal world and the empirical world requires the concept of **noumenal mind**, and we arrive at this through the explanation of illusions in the empirical world. Recall that empirical reality was defined as everything empirical that is non-illusory (54); so illusions, although empirically actual, are empirically unreal — where the empirically actual is all that is perceptible, which is the empirical world. Which is to say that not all of the actual empirical world is real because some of it is illusory.

In the middle of the twentieth century most Anglophone philosophers dismissed illusions. They were deluded by Ludwig Wittgenstein (1889-1951) into believing that all philosophic problems could be either solved, or else shown to be pseudo-problems, by "analysing the rules of the language game" in which these problems arose. (The supposition behind this, in the language of the present book, is that all philosophical problems have nominal meaning only.) One such problem was the problem of illusion, which is the problem that any explanation of illusions makes them mis-representations, or false images, of reality, in the brain; but we experience them as outside our heads and so not in the brain. This paradox was then linguistically "resolved" in two ways. One was by conflating illusion with delusion and claiming that

if an illusion did not delude us then it was not an illusion —
and since no illusions did delude us, there are no illusions. The
other was to say of an illusion that "I see reality but I see it as
illusory" or "I see reality but it appears to be illusory" — a
distinction that led to a ludicrous controversy as to whether
illusions were adverbial or adjectival. By these sophistical
means the problem of illusion was swept under the carpet and
naive realism — the belief that the empirical world is reality
— has dominated philosophy (and physics) ever since. But the
problem of explaining illusions is still a problem needing a
solution, and naive realism is still naive.

Instead of claiming that illusions are not really
illusions, or that philosophers may safely ignore them, we
need an explanation of why there are illusions at all. And there
is only one satisfactory explanation: illusions are
misrepresentations of reality — false images of reality —
rather than reality itself. The empirical spoon, half immersed
in a glass of water and seen from above, is bent at the surface,
but the real spoon is not bent — as we can discover by running
a finger down it. So the bent spoon is an image of the real
spoon, the empirical spoon is bent, so the empirical spoon is
an image of the real spoon, which is not bent, so the empirical
spoon is not the real spoon.

An ancient argument, called the argument from
illusion, goes: there are no empirical intrinsic differences
between illusions and non-illusions, so if illusions are images
of reality then so must non-illusions also be images of reality.
Illusions are false images of reality and non-illusions are true
images of reality. Hence everything empirical is an image of
reality, not reality itself. This Platonic conclusion is a denial of
naive realism.

From what we have discovered so far, "reality itself",
of which the empirical world is an image, is the noumenal
world of theoretical physics and metaphysics.

The argument from illusion has never been refuted;
illusions are an indisputable fact; and yet hardly any scientist,

mathematician, or philosopher today accepts the conclusion of the argument from illusion. Most of them with whom I have discussed this problem are as passionately closed-minded about it as are creationists closed minded about evolution. Their common sense reasoning is simply that brain images are inside the perceiver's head and their originals, the real objects, are outside the head; and what we perceive is outside our heads, hence real. But some of what we perceive is illusory and so images inside our heads, even though it is perceived as outside our heads. So we have to resolve the paradox, the contradiction between common sense realism and the argument from illusion: are illusions outside out heads or inside? Or both? But how can they be both?

Let us first consider some examples. Visual space is illusory: all visual magnitudes, in all three dimensions and in all directions, diminish with distance from the observer. Imagine, or look at, a straight road with equally spaced trees along it. With distance the road gets narrower, the trees get smaller, and the trees get closer together. Or railway tracks seem to meet in the distance, and a train going away down them appears to get smaller and shorter as it goes. Until the Renaissance, artists were unable to render these contractions properly. It was the invention of projective geometry that enabled them to do so, with perspective drawing. Visual space is described by projective geometry, while real (local, terrestrial) space is described by Euclidean geometry. In projective geometry parallel lines meet in the distance, in Euclidean geometry they do not meet. In empirical visual space parallel lines meet in the distance, in real space they do not. The scientific explanation of this is, of course, well known: images on the retinas of the eyes diminish in size as their original objects get farther away. The objects do not diminish with distance, but the images do so diminish; the objects are real, the images are, indisputably, images; the objects are noumenal, the images are empirical. What we see is images of reality, because it all gets smaller with distance;

but yet it is reality because it is outside our heads. The bizarreness of visual space is made worse by the fact that every observer occupies a personal *here and now* which is the co-ordinate origin of his or her personal visual space, so that this personal space is characteristically different from every one else's visual space — and since qualitative difference entails quantitative difference — Q,Q. (70) — everyone has their own numerically distinct visual space, disjoint from every one else's and completely subjective. Anyone inclined to dismiss all this out of hand should consider the question: how far away from your eyes must a visible object be in order for you to see its real size?

Another example comes with microscopes. If you see a bacterium with an optical microscope, or a virus with an electron microscope, or an array of molecules with a scanning tunnelling microscope, does the microscope enlarge the object, or enlarge an image of the object? Obviously, it does not enlarge the object — how could it? — so it enlarges an image of the object. What you see is enlarged and hence is the image of the object, not the object itself, which is not enlarged. The object is real, noumenal, and the image is empirical. You can *never* see a bacterium because it is too small; but you can see an enlarged image of it. And the bacterium and the image of it are qualitatively different hence Q.Q. again applies: they cannot be identical, numerically one, because they are different sizes.

This argument can be generalised to: everything seen through a lens is an image of reality, not reality itself; our eyes have lenses; therefore everything seen is an image of reality, not reality itself, and so is in the brain. This is also true of telescopes, without which we would have no empirical astronomy or cosmology. And all this can be generalised from the visual to all the other senses: everything perceived by means of a sense organ is an image of reality, not reality itself — and is in the brain. Which is to say, everything empirical is an image, true or illusory, of noumenal reality. But yet everything empirical (except headaches, and tastes and smells) is experienced as external to our heads, hence real according to common sense.

Another ancient problem is that of **secondary qualities**, which are manufactured in the brain of the perceiver as a result of receiving signals from the sense organs, via the afferent nerves. Thus electromagnetic radiation, reflected off noumenal grass, lands on retinas as an image and is transduced into an image composed of sensations of green — and this green is a secondary quality, in the brain of the perceiver. As soon as it is understood that this green sensation is manufactured in the brain then it follows that the noumenal grass is not green; instead, theoretically, it contains chlorophyll, which reflects the electromagnetic radiation of the frequency that generates the green sensation. Again, acoustical vibrations in the noumenal air land on to ear-drums and are transduced into sounds — and these sounds are secondary qualities. And similarly for all other concrete qualities: tactile sensations, tastes, and smells. As such, all secondary qualities are illusory, hence unreal — where the real is defined as all that exists independently of perception. It is indisputable that secondary qualities are images of their originals, hence all concrete qualities are images of reality, in our brains, not reality itself. But empirical concrete qualities are experienced as being outside our brains, hence are real.

Again, consider things seen out of focus. You can focus your eyes on something nearby and then move them to more distant objects without changing the focus, and the new objects are out of focus. Or you can wear someone else's glasses, so that everything seen is out of focus. It is impossible for reality to be out of focus; images, and only images, may be out of focus. So everything that *may* be out of focus *must* be images, *cannot* be reality.

Last, stereoscopic vision requires two two-dimensional images, out of which a three-dimensional empirical world is constructed. Since we see stereoscopically, the empirical world must consist of images.

There are a number of attempted defences of naive realism — the belief that what we perceive is reality, not images of reality — which should be mentioned here, if only to refute them. One, which goes back to Aristotle, is that we perceive real objects *by means of* images in our heads. This would work if the *means* could be spelled out; but they never have been — and cannot be so because this explanation will soon be shown to be false. Another, quite popular these days, is the doctrine of indirect perception. It is said that to perceive an effect is to perceive its cause, indirectly. So to perceive an image is to indirectly perceive its cause, the real object, and since indirect perception is perception, we thereby perceive reality. The hollowness of this is clear as soon as it is pointed out that indirect perception is a synonym for belief: when we supposedly indirectly perceive the cause of an empirical object, we do not perceive that cause, we only believe in it. For example, when we perceive the Sun we believe we are seeing the real sun, although we are only perceiving an image of it — as is proved by the possibility of the real Sun ceasing to exist, but the image of it continuing to exist for eight minutes after the cessation. Belief is a perception substitute, which we invoke as an alternative when perception fails, and belief is false more often than not. A third defence of naive realism is the claim that we can perceive down causal chains, from empirical effect to real cause, much like "seeing down" a microscope, so that by perceiving an image we perceive reality. But if this were so then we would be able to see the Big Bang, or, perhaps, God. A fourth defence is a doctrine of projection: we supposedly project secondary qualities and illusions out from our brains on to real objects outside. This suffers from the same problem as Aristotle's "means": unless this projection can be explained, it achieves nothing. And it cannot be explained: we do not project as movies are projected on to a screen in a cinema (we would need a lamp in our heads in order to do so), we do not project as a slingshot or gun projects, we do not project as shadows are projected on to the ground, we do not project geometrically — and there are no other kinds of projection. To suppose that there is yet another kind, without explaining how it works, is futile.

However, most naive realists do not defend their position, they are simply dogmatic about it: we perceive reality, the empirical world *is* reality, period. Any alternative is too repugnant to consider.

So what has gone wrong, and how can it be sorted out? I have to say that although the solution is very simple logically, it is very difficult psychologically; very, *very*, difficult; so difficult that most people, on hearing the solution, immediately, passionately, deny it. It is only clear thinkers who can accept the logic of it.

The problem was probably solved by Socrates (209); it was first explicitly solved by Leibniz[19] (214), but not appreciated by other philosophers. John Locke (1632-1704) and Bishop Berkeley (1685-1753) came close to solving it but retreated into common sense. The solution was rediscovered by Bertrand Russell[20] (1872-1970), who acknowledged Leibniz' priority, and the solution was later rediscovered by myself during a three-year stay in the Canadian Arctic (a great place to think philosophically). I have since called it the Leibniz-Russell solution.

First, let us state the problem as starkly as possible. According to the above arguments, everything we perceive — everything empirical, or known through the senses — is an image of reality, not reality itself. If we ask *where* these images are, the only sensible answer is that they are in our brains, brought there from the various sense organs by the

---

[19] Leibniz, "Monadology".

[20] Bertrand Russell, "Human Knowledge, Its Scope and Limits", Part 3, Allen and Unwin, London, 1948.

afferent nervous system. Your empirical qualities are concrete images in your brain, empirical objects are structures of empirical qualities, and your empirical world is a structure of empirical objects. So this means that your entire empirical world is inside your head. No wonder common sense reacts so negatively! So is your empirical world outside your head, as experience tells you, or inside your head? Or both?

Logically, this problem is equivalent to reconciling the two statements "The triangle is inside the square" and "The triangle is outside the square", and this can only be done by denying one or both definite articles: there are two triangles, or two squares, or both, as in Fig. 11.1.

The logic of the Leibniz-Russell solution is that if *everything* empirical is an image of something noumenal — an image of reality, not reality itself — then your own empirical, naked, body is an image of your real, noumenal body. Your

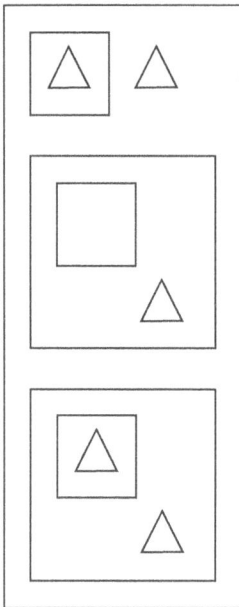

Fig. 11.1

empirical body is an empirical object, like all empirical objects, and consists of empirical sensations — except that it is unusual in consisting as well of extra, internal, sensations, unlike all other empirical objects. More particularly, your empirical head is an image of your noumenal head. Your noumenal head contains your noumenal brain, which contains images of the noumenal world, which images constitute your empirical world, which empirical world contains your empirical body. Thus the empirical world is outside your empirical head and inside your noumenal head. Very simple, logically. Clearly, in solving this problem two heads are better than one.

Putting this solution another way, if you go outside on a sunny day then beyond the empirical blue sky is the

178

inner surface of your noumenal skull. This highly disturbing conclusion is hereafter abbreviated to S.B.S., for 'skull beyond sky'.

Even more disturbing is the subsequent necessary denial of three common sense metaphysical beliefs: the belief that empirical objects continue to exist when unperceived, the belief that other empirical people have minds, and the belief that there is only one empirical world, with all of us in it. Being images, empirical objects exist only for as long as they are perceived: *esse est percipi* (to be is to be perceived), as Bishop Berkeley put it — just as images on your television screen exist while your set is switched on and cease to exist when your set is switched off. And, secondly, being images, other empirical people do not have minds — how could an image have a mind? — they only behave as if they have minds because they are images of noumenal other people, who do have minds. Equally, people seen in movies or on television do not have minds, although they behave as if they do because they are images of noumenal people, who do have minds. And thirdly, there are as many empirical worlds as there are perceivers.

Needless to say, you do not have to think about S.B.S. while doing everyday tasks, any more than you worry about the Earth hurtling at 30,000 k.p.h. (18,000 m.p.h.) around the Sun. (Such a dangerous speed! What if something was sitting in the way?) But to ignore the Leibniz-Russell solution in favour of common sense, while trying to understand what 'underlying' means in the statement that theoretical science explains empirical phenomena by describing their underlying causes — that is to be deeply in error.

We may note in passing that the noumenal is larger than empirical images of it. You can clutch your empirical head with your empirical hands, to find it to be about twenty empirical centimetres across. But your noumenal head contains the empirical blue sky, which is more than twenty empirical centimetres away. Note also that the empirical stars are the same distance away as the blue sky, which is a hemisphere having a radius equal to the distance to the geographical horizon from your personal "here and now"; it is the noumenal stars that are light-years away. The empirical stars are mere points of light in the empirical black night sky, images of the noumenal stars, which are suns.

It is also worth remarking that this is not the first time common sense has been shown to be wrong, with the result of much passionate debate. It happened with Copernicus' (1473-1543) denial of geocentrism (The Earth moves!), and it happened again with Darwin (1809-1882) and the theory of evolution (Such vile atheism!). It would also have happened with Einstein's two theories of relativity and with the standard model in quantum mechanics were it not for the fact that these lie so far outside common sense that the only result is a little cognitive dissonance among the cognoscenti.

So we now see why the relationship between the noumenal world and the empirical world requires this concept of **noumenal mind**, which is an emergent out of noumenal brain, and which contains the empirical world; that is, the empirical world is noumenally ideal, it consists of noumenal ideas which are images of the noumenal world. And for this we need to explain how the noumenal mind is conscious, and how it can act. Such explanations are not difficult, once the fundamental nature of relations, and S.B.S., are taken into account.

We are going to assume a theoretical principle of noumenal mind, called the **mind hekergy principle**, which is the principle that the noumenal mind increases its hekergy whenever possible, and if this is not possible, it works to maintain its hekergy whenever possible, and if this is not possible, it works to minimise its loss of hekergy. If this last fails the result is death. This principle originated with Erwin

Schrodinger[21] (1887-1961), who first defined life as very high negative entropy in dynamic equilibrium. (Note that 'negative entropy' is a misnomer: it should be 'negated entropy', entropy with a minus sign in front of it, because in physics there is no negative entropy because there is no temperature below absolute zero; note also that hekergy is negated entropy generalised to relations.) The mind hekergy principle explains creativity and our desire for the valuable, and is quite plausible given that evolution is a hekergy increasing process.

We know that the noumenal human brain consists of about 200 billion neural cells, arranged together into the most complex structure known to us. Since mind is emergent out of this, it is reasonable to assume that the most elementary part of a noumenal mind is one of the most elementary parts of a noumenal brain, which is a neural switching: an *on* or an *off*, where an *on* is a relation between two neural cells and an *off* is a relation dissimilar to an *on*. An *on* is an **atomic noumenal idea**, as is an *off*, and other **noumenal ideas** are structures of these, structures of structures, and so on up — as in the Grand Structure. Thus a transient structure of neural switchings is a **noumenal sensation**, brought into the noumenal brain by the afferent nervous system, from various sense organs; it will be called a **mid-sensation** because it is causally midway between a noumenal property in the noumenal world and a sensation in the empirical world. A structure of mid-sensations is then a transient **mid-object** and a structure of mid-objects is a transient **mid-world**. Our aim is to show how these become causes of some of the content of consciousness, in the form of empirical sensations, empirical objects, and empirical world. This requires a subject of consciousness, which is the **noumenal ego**.

From a subjective point of view, if anyone asks "What am I?" the most reasonable answer is that I am at least all of

---

[21] Erwin Schrodinger, "What is Life?", Cambridge University Press, 1944.

my memories and beliefs, for these are what distinguish me, qualitatively, from everyone else; I am also a conscious being — conscious of the empirical world, of my own body in the empirical world, and conscious of my own mind. So let us assume that the noumenal mind is able to map transient mid-sensations, mid-objects, and mid-worlds into permanent sensations, objects and worlds; these are **mid-memories**, or **noumenal memories**. A structure of mid-memories then becomes the noumenal ego, following a principle of movement and configuration of mid-memories, called the **L.A.L. principle**, where L.A.L. stands for 'like attracts like and repels unlike.' This is formulated by analogy with principles of force in physics: namely, an inverse square law of like-attracts-like or else like-attracts-unlike-and-repels-like, as in forces in gravitation, magnetism and static electricity. In the noumenal mind, noumenal ideas attract likes and repel unlikes; but there is in this another factor, in that the degree of likeness, or degree of similarity, is also a parameter.

As we saw earlier (39), given any two relations, if they have a number, s, of similar properties and a number, d, of dissimilar properties, and one has m more properties than the other, then the **degree of similarity**, L, between them is $L=s/(s+d+m)$, and the **degree of dissimilarity**, D, between them is $D=(d+m)/(s+d+m)$; and if $D=1$ then $L=s=0$, and if $L=1$ then $D=d=m=0$, and each varies between 0 and 1. Thus if we have a degree of similarity L, then $1-L$ is the degree of dissimilarity. These definitions of degree of similarity and degree of dissimilarity may be extended from those of single relations to those of structures by defining corresponding terms of the relations and their configurations, and the degrees of similarity or dissimilarity of these, and of other relations between the terms, and repeating this for lower and lower levels, down, if necessary, to axiom relations, where an axiom relation in a noumenal mind is an atomic noumenal idea, an *on* or an *off*. The degrees of similarity at each level could then be averaged, and the averages for different levels weighted to form an overall average, with the weights decreasing with level. If desired the weights for a given level and all lower levels, could be made zero, for convenience.

We then define the force between two noumenal ideas as proportional to:

$$\frac{(L - 1/2)(H_1 \times H_2)}{d^2}$$

In this formula L is the degree of similarity, between the two noumenal ideas. If L is greater than ½ then the force is positive and, by convention, thereby an attraction; while if L is less than ½, a negative force, and so repulsion, results. (In electrostatics and in magnetism $(L-½)$ in effect has only two possible values: -1 and $+1$; and in gravitation $(L-½)$ in effect has only one value, $+1$.) H_1 and H_2 are the total hekergies of each of the two ideas, and d is the distance between them. An inverse square law is assumed not only because other laws of force obey it, but because it is mathematically necessary in a three-dimensional space.

All the noumenal memories that form the ego have a major component in common, which is the mid-memory of the mid-body of the person concerned, the mid-body being an image of her or his noumenal body, of which latter the noumenal brain is a part. This common component makes these mid-memories largely similar, so that they mutually attract, by L.A.L., and form a structure which is the **noumenal ego**. The top idea of this ego might be called the Self. Not only does the ego grow with new memories, but it later also grows with beliefs, as will be explained shortly (188).

If now a noumenal idea comes close to the ego then L.A.L. forces between this idea and the noumenal ego will produce distortions in the ego, much as the Moon produces tides on Earth, but much more complex because of the greater complexity of the ego and of the noumenal idea. These distortions — various degrees of pushes, pulls, bends, turns, twists, etc. — of the noumenal ego are its **consciousness**. A mid-sensation therefore produces consciousness that is an

empirical sensation, while a mid-object produces consciousness that is an **empirical object**, and a mid-world in turn produces consciousness that is an **empirical world**. All of these contents of consciousness are transient, but mid-memories of them are formed and may later produce consciousnesses that are **empirical memories**.

Notice that this is an explanation of consciousness of empirical data, in that it describes the underlying, or noumenal, causes of those data. It also requires that all the contents of consciousness are within the noumenal ego. In other words, if you go outside on a sunny day then beyond the empirical blue sky are the outer parts of your noumenal ego, beyond which is your noumenal skull.

Another point about the contents of the consciousness of the noumenal ego is that they consist entirely of concrete qualities and relations between them. The concrete, it will be recalled, was defined as any empirical sensation: colours, sounds, tactile sensations, smells, or tastes, also known as secondary qualities; and the abstract was defined as anything lacking concrete qualities, and was declared to be any relation or property of a relation. (Empirical objects are structures — relations — of empirical qualities.) And although the concrete sensations have value in their own right — beautiful colours, wonderful sounds, gorgeous tastes, etc. — their importance is that by acting as terms of empirical relations they enable us to perceive these empirical relations; and these empirical relations are fairly reliable copies of noumenal relations. It is the noumenal relations that we need to know about, in order to survive. For example, we can perceive various shapes of green which we call leaves, and know that this leaf is good to eat and that leaf is poisonous; or that that branch is thick enough to bear our weight but that other one is not. The survival that such perception of relations enables is survival of the noumenal body, which is empirically perceived as survival of the empirical body. Since survival is our first imperative, perception of empirical relations is essential. And we can easily learn new such relations: learning the magnitude of the relation between speed and distance when driving a car has survival value.

Note also that empirically emergent concrete qualities are noumenally emergent abstract relations, because they are alterations to the structure of the noumenal ego. As was remarked earlier (40), all emergence is emergence of relations, although it does seem as if concrete qualities emerge — as with the emergence of colours in the sky at dawn.

Not only is the noumenal ego conscious, it is also able to act. We may suppose that some noumenal ideas, called **noumenal motor ideas**, instead of being transmitted by the noumenal afferent nerves from sense organ to brain, go from the brain by the noumenal efferent nerves in order to move noumenal muscles. A new born or young ego inherits some of these, such as its ability to suck and to smile, and it can learn to direct other motor ideas by learning correlations between willing to move them and getting results in the form of movement of its own empirical body. Although the ego may later think of this as willing its empirical body to move, what is happening is that it is willing its noumenal body to move, and this noumenal movement is the cause of the movement of its empirical body, since its empirical body is an image of its noumenal body.

A particular form of action is language, in speech and writing. A combination of sounds may be bonded to something in the empirical world of the ego — a man, a horse, or a dog, for example — so that the ego may point its empirical finger to an empirical man and ostensively define the word 'man'. As any parent knows, this is how young children learn to talk. Concepts follow soon after — young children soon learn to count, for instance. Writing follows later, with the bonding of structures of letters or symbols, or of ideograms, to vocal words, to images in the imagination, and to abstract ideas in thought.

With speech, one ego may talk with another. One, using motor ideas, produces noumenal sounds — acoustical vibrations — in the noumenal world which reach the noumenal ears of the other and produce noumenal acoustical sensations in the noumenal brain of the other, which are mid-sounds which produce empirical words which produce meaning — relational, extensional, nominal or other meaning — by means of the ideas bonded to similar words in the mind of the listener. Thus the second ego gains the meanings intended by the first. The second ego can respond in the same way. This is a rather more complicated process than common sense would have it, but one that explains all the empirical facts, including the fact that acoustical vibrations and heard sounds are two, not identical. Remember Bishop Berkeley's puzzle about the tree that falls in the forest when there is no one around to hear it: does it cause a sound, or not? It does cause a noumenal sound, but does not cause an empirical sound.

Noumenal people may compare their empirical worlds and discover that they are very similar to each other, especially when their viewpoints are close together. This leads to the assumption that all these people are in one and the same, or identical, world; and that this empirical world is real, in the sense that empirical objects continue to exist when unperceived. And because of communication each person assumes that other empirical people have minds more or less similar to their own. These three metaphysical beliefs of common sense are all false: they are oversimplifications. Each conscious person has their own, numerically distinct, empirical world because it is a numerically distinct image within a numerically distinct noumenal mind, within a numerically distinct noumenal skull; these distinct empirical worlds may be largely similar, but they cannot be identical: noumenal brains cannot intersect because noumenal skulls cannot intersect. This is an interesting point because, it will be recalled (70), *similarity* is characteristic of relational meaning while *identity* is characteristic of extensional meaning: thus on this question naive realism (165) has extensional meaning while S.B.S. has relational meaning. Similarly, empirical objects are somewhat *similar* to noumenal objects — the former are images of the latter, after all — but not thereby *identical* with them, and so not real; and empirical other people are *similar* to noumenal people but not *identical* with them, and so not conscious. Relational meaning is abstract and difficult, while extensional meaning is usually concrete and easy; which is why common sense prefers extensional meaning, in spite of all the errors that it introduces. The errors are introduced through ignoring the logical principle that qualitative difference entails quantitative difference (70); ignoring this principle was called the identity error (69). For example, "My empirical world is largely similar to yours therefore they are numerically one, identical".

Another common sense error of over simplification is common sense explanation, in which perception of causes is assumed, falsely. We earlier (62) distinguished relational necessity and extensional necessity, the first being singular possibility and the second being universality; and relational necessity necessitates universality, but not *vice versa*. No one can perceive universality, but it is easy to interpret high correlation as universality, by induction (107), and then to interpret this pseudo-universality as singular possibility. This pseudo-necessity is then assumed to be a causation, its antecedent being a cause and its consequent being an effect, so that the effect is explained by citing the pseudo-cause. For example, there is a high correlation between flashing of lightning and sound of thunder, and it is believed by many that lightning causes thunder. But it does not, because they are both caused by an atmospheric electric discharge, and so are correlated. There are even those who believe that a flash of lightning and an atmospheric electric discharge are identical, even though this is impossible because qualitative difference necessitates quantitative difference. Even worse than such false explanations are explanations based on low correlations, such as astrological explanations. Such over-simplified common sense explanation is perhaps acceptable for everyday living but not for scientific explanation.

It was said earlier that the empirical and noumenal are disjoint (7), but there is an exception. This disjointness is important in clearing the mind of common sense errors, such as the belief that objects seen in microscopes are empirical and real, rather than being images of the noumenal and so numerically distinct from the objects. The one exception is that, in the final analysis, the empirical world of each person (including their empirical mind, known introspectively) is within their noumenal ego and so part of their noumenal mind, hence part of the noumenal world. The two worlds, noumenal and empirical, are disjoint only in that the noumenal-world-less-empirical-world-and-less-empirical-mind is never empirical. So the noumenal is all that exists, all that is actual, and each person's empirical world is a special part of it.

We finish this chapter with a miscellany of explanations of psychological facts, with the goal of demonstrating that the present theory of mind has reasonable scope and density of detail — two of the criteria of good explanation. They are arranged alphabetically, and since there is considerable cross-referencing between them, words that refer to other entries in the list are in italic typeface.

Abstraction is a special kind of *mapping*, as are *discriminatory mappings*.

Attention. The ego can distort itself in order to focus on what is of interest to its *selfish goals*. This focussing is its attention.

Agent. Anything that has some awareness of its environment, some control over it, and goals within it is an agent: so the noumenal ego is an agent, as is the *oge*.

Belief. The difference between the ego considering a proposition and believing it is that in considering it, it is simply conscious of it and thinking about it, while in believing it, the ego incorporates the belief into itself, in similar fashion to a memory. It then, for the ego, becomes *my* belief.

Bonding. Two or more noumenal ideas can become permanently joined to each other, as in the bonding of an abstract idea to a word to form a concept, or in the bonding of a motor idea with the ego's intention of a particular muscular movement.

Classification. It was claimed earlier that we are born to classify (36). This fact can be explained by L.A.L, (ideas attract like ideas). 'Like' means similar, and similarity is the basis of classification because classes have intensions based on similarity: whatever X may be, $\{\aleph X\}$ is the class of everything similar to X and $\aleph X$ is the intension of this class, because of X having $\aleph$ as an upper. (Things and qualities, as well as relations, may have uppers, because they may be terms of relations.) This explanation suggests that the word class is to be preferred over the word set because it derives from 'classification'.

Contents of the empirical head. Apart from an occasional headache, and tastes and smells, the empirical head contains nothing; it is the noumenal head that contains a brain and mind. The empirical head may contain an empirical brain, but only if the head is opened to reveal it (see *esse est percipi* below). That is, if the empirical head is opened, this opening is an image of the opening of the noumenal head, exposing the noumenal brain which is imaged as an empirical brain in the opening of the empirical head. See also *Russell's joke.*

Creativity is increase of hekergy, in the empirical mind or in the empirical world, each of which is a consciousness of hekergy increase in the noumenal mind, which consciousness is an image of a hekergy increase in the noumenal mind or noumenal world.

Discrimination. We may suppose that the ego is capable of special *mappings* of empirical data that enable it to discriminate. For example, a boundary is a series of contiguous dissimilarities; if this is mapped, the result is a shape. A scale mapping, up or down to similarity, gives relative size. And a mapping of a small patch gives a colour or a tactile sensation. Such discrimination evolved because of its survival value.

Esse est percipi. According to Bishop Berkeley, to be is to be perceived; which implies that everything empirical exists only as long as it is perceived. According to the present theory he was quite correct in this, in that the empirical world consists of transient images only, but Berkeley failed to draw the logical conclusion of the existence of a noumenal world; instead he fell back on to naive realism and declared that empirical objects continue to exist between occasions of us perceiving them because they are perceived by God.

External world. The external world is the world external to the body of a perceiver. The noumenal world is external to the noumenal perceiver, and the empirical world is external to the empirical perceiver but internal to the ego of the noumenal perceiver. If the old philosophical question "Does the external world exist?" is raised, it is important to specify which external world is intended; to a naive realist the question appears ludicrous.

Feeling. Consciousness of hekergies; see also *thought.*

Goal. A goal, of an *agent*, is something that will increase the hekergy of that agent.

Intuition. The ego may become aware of other parts of the noumenal mind and in doing so becomes conscious of some of their content. Two such parts are the *psychohelios* and the *oge*. Intuiting the former gives solutions to problems, and intuiting the latter gives revelations and moral certitude. *Prejudice* is easily misunderstood to be one or the other of these.

Irrational. The irrational is any assemblage of noumenal ideas that are arranged by L.A.L. (like-attracts-like-and-repels-unlike), as opposed to the *rational*. A young ego is entirely irrational, while a mature ego may become more rational through its desire to increase hekergy, in accordance with the mind hekergy principle.

Love. A willingness, by an *agent* to give unconditionally to its beloved. This is only possible, in the case of the ego, if the ego incorporates the loved mid-person into the ego, so that loving it is an ego hekergy increase. Such incorporation is *bonding*, in two senses of the word.

Mapping. Mapping in the noumenal mind is a copying of ideas, in whole or in part. Mapping part of an idea of a woman and part of the idea of a fish, and *bonding* the two together gives the idea of a mermaid. A mapping is a relation.

Moon. An old puzzle about the Moon is that in perceiving it we are conscious of both the Moon and its distance; our consciousness seems to extend from our empirical heads out for about four light-seconds — between 200,000 and 400,000 km. — to the Moon; but how can it do that? Noumenally, one's own empirical body, the empirical Moon, and the empirical distance between them, are all within the noumenal ego; the four light-second distance is noumenal, not empirical; the empirical distance of the Moon is the distance of the empirical black night sky from the empirical body, about 8km. in the projective geometry of visual space.

Oge. This is a second *agent* in the noumenal mind, discussed in Chapter 12.

Perception. There are two kinds of perception: **noumenal perception** is a causal process of producing, in a noumenal ego, noumenal images of noumenal objects; and **empirical perception** is the last stage of noumenal perception, experienced simply as a given. Noumenal perception is theoretical perception, and so explanation of empirical perception.

Pleasure and pain. Pleasure is consciousness of hekergy increase, pain is consciousness of hekergy decrease. These may be physical, due to hekergy changes in the noumenal body which are mapped into the empirical body; or mental, due to hekergy changes in the noumenal mind which are mapped into consciousness. Pleasures may be increases in the empirical world, such as increase of wealth or fame, increases in the empirical body such as sensual pleasures, or increases in the empirical mind such as a result of *creativity* or of learning. Pains may be corresponding losses, of which a particular kind is loss of belief: a *belief* is part of the ego and to lose a belief is a diminution of the ego — as in losing the belief in naive realism: it is acceptable only if replaced by a better belief, such as S.B.S.

Prejudice. A prejudice is a *belief* that is strong enough to select evidence in its favour, and reject evidence against it, by L.A.L. Because of this the prejudice is, to the holder of it, obviously well substantiated, so obviously true. One can discover one's own prejudices by searching for extremism among one's beliefs — particularly beliefs in politics or religion.

Psychohelios. The psychohelios is a structure in the *noumenal brain* that is wholly *rational*. It is discussed in Chapter 13.

Rational. Noumenal ideas that are arranged with maximum hekergy are rationally arranged, as opposed to ideas that are arranged by L.A.L., which are *irrationally* arranged. Maximum hekergy occurs when ln.t/e is a maximum, meaning that $e=1$; and $e=1$ is a singular possibility (7, 42, 43), which is a necessity; and necessity in thought is logical, rational. And maximum hekergy is maximum value.

Recognition. If a mid-object attracts a mid-memory to itself because they are *similar* this results in recognition: mid-recognition and empirical recognition.

Russell's joke. Bertrand Russell, who had a puckish sense of humour, said that when a brain surgeon sees another man's brain, what he sees is in his own brain, not in the other man's brain. He meant that what the surgeon sees — the empirical other man's empirical brain — is in his own noumenal head, not in the other man's noumenal head, although it is in the empirical other man's empirical head. Or, more clearly, if the brain surgeon is a woman, when she sees a man's brain what she sees is in her own noumenal head, not in the man's noumenal head — although it is in the man's empirical head in her empirical world.

Self. The essential self is an emergent noumenal idea that unifies the collection of mid-memories of the noumenal body, and of *beliefs*, into a whole, a whole greater than the sum of its parts, a whole which is the noumenal ego. The self is the top relation of the whole. The ego can be vaguely aware of this idea of the self.

Selfishness. The noumenal ego needs to increase its own hekergy, in accordance with the mind hekergy principle; this need is the basic attitude of the ego, and is called selfishness.

Thought and feeling. Thought is consciousness and manipulation of ideas according to their meanings. *Feeling* is consciousness and manipulation of ideas according to their hekergies, which are experienced in consciousness as subjective *values*.

Value. Value is hekergy. Absolute values are noumenal hekergies, relative values are subjectively distorted consciousnesses of these — distorted by the irrationality of the ego.

Chapter 12. Social Relations and Conflict.

Mid-memories of the mid-person do not always go to augment the ego because they usually include mid-memories of other mid-people. These other mid-people may be approving of the mid-person of the ego, or disapproving. If they are approving then the approval makes them like — similar to — the ego, and so their memories are attracted to the ego, by L.A.L, while if they are disapproving then the memories are unlike the ego and so repelled from it. All those that are unlike the ego are thereby mutually attractive, and so they form a structure, a complex, in the noumenal mind, second to the ego. Like the ego, this complex is an agent (188), something that has a goal in, some awareness of, and some control over, its environment. Based on disapproval of the person of the ego, it is basically hostile to the ego; it is like an anti-ego, and so is called the **oge**, a word which is 'ego' spelled backwards, and pronounced to rhyme with fogey. The concept of oge has considerable explanatory power.

Ego-approving mid-people are primarily mid-parents, who approve of their child because they love (191) it. But sometimes they disapprove, which leads to a brief withdrawal of love, and so they send mid-memories to the oge. The proportion of love to the absence of it is crucial to the mental health of the child and later adult, since mental health includes a balance between ego and oge.

We saw that the ego is basically selfish because it needs to increase its own hekergy, in accordance with the mind-hekergy principle. Similarly, the oge needs to increase its own hekergy, and because it represents other people it is concerned with the good of the society of the ego; so it is basically **moral**; it is opposed to the selfishness of the ego. As such it is the source of our moral sense and of feelings of guilt, shame, and duty; and it is also a necessary condition for internal conflict.

There are five conditions for conflict, each necessary and all jointly sufficient: they are (i) two agents, (ii) with some of their situations in common, (iii) each with goals, incompatible with the other's goals, in that common situation, (iv) each with some awareness of that situation, and (v) each with some control over that situation. The gaining of one agent's goal is defeat of that agent's opponent. The agents may also be teams of agents. Familiar empirical conflicts are games such as chess or poker, sports such as tennis or football, political and industrial conflict, and war. Internal conflicts are inclination/duty conflict and neurotic conflict.

The ego and the oge are quite different with respect to their shapes. The ego may be thought of as roughly spherical, as L.A.L. brings mid-memories to one central mid-idea. But the oge consists of mid-memories of many different mid-people. These may be thought of as oge-people, or mid-people, such as mid-mother, mid-father, mid-family, mid-friends, and mid-strangers; and, usually by repute, oge-enemies such as oge-criminals, oge-enemy-soldiers, and oge-terrorists. ('Oge-' is a special case of 'Mid-' in this context.) They will be strung out in a line, their nearness to the ego being according to their degree of approval or disapproval of the ego, thus making the oge sausage shaped. But shape difference does not stop there, because the ego is polarised as well as spherical. The ego contains two prejudices, of a kind which C. G. Jung (1875-1961) called archetypes of the collective unconscious[22]: the *persona* and the *umbra*. The persona (mask) is all that the ego wants others to believe of it, and the umbra (shadow) is all of it that it does not want others to know about it. The first desire is based on the need for approval, and the second on the need to avoid disapproval. For unknown reasons the persona is at the top of the ego and the umbra at the bottom. So the sausage-

---

[22] Jung supposed that there is only one collective unconscious, to which all people had access. In the present context this is implausible; it is better to suppose that some contents of unconscious minds are similar among all people. To infer identity from these similarities is to commit the identity error.

shaped oge bends longitudinally around the ego, with the oge-parents attracted to the *persona* at the top of the ego and the oge-enemy attracted to the *umbra* at the bottom; they are so attracted because the characteristic of the oge-lover is love and the characteristic of the oge-enemy is not only absence of love, but hate and evil. Once so arranged oge-people spread around the ego, as a concentric shell, with oge-strangers at the equator. If the ego can sense the presence of the oge — some people can and some cannot, to varying degrees, as we shall see — the oge-lover will be above the empirical blue sky on a sunny day, or the black sky of night, and the oge-enemy will be below the empirical ground: the traditional locations of heaven and hell, in other words. Note that these locations are in the noumenal mind; they are neither astronomically above and below the noumenal Earth, nor in the empirical world. Thus the oge-lover is one meaning of the word God and the oge-enemy is one meaning of the word Satan, as is further discussed in Chapter 13; the equator of the oge is then purgatory.

Besides shape, an important difference between the ego and the oge is their relative strength. To varying degrees, in different people, the ego may dominate the oge or the oge may dominate the ego. An ego-dominant type is more selfish than moral, while an oge-dominant type is more moral than selfish. Various behavioural characteristics, such as duty, public spirit, conventionality, loyalty, obsequiousness, conscientiousness, political correctness, obedience of moral proscriptions and prescriptions and of orders issued by other people, charity, sensitivity to the judgements of other people, shunning of responsibility, susceptibility to embarrassment and shame, and being in favour of the under-privileged and hostile to the over-privileged: such are characteristic of the oge-dominant, in various ways and to various degrees. And the opposite of these are characteristic of the ego-dominant: concentration on selfish ambition, indifference to the opinions of others, to embarrassment, to shame, and to taboos, etc. In organisations where everyone is ego-oge unbalanced to some extent, the result is an authoritarian social group, a hierarchy: an aristocratic society, the military, the civil-service, a university, a church organization, a large business, a political party, a trade union — all have order-givers at the top and order-receivers at the bottom. Such people have a range of rudeness-ability: they can be rude to those below them but not to those above them — as with the peck-order among chickens. Such authoritarian groups are notorious for their internal political scheming: a matter of ego-oge conflict as well as conflict between individuals.

If the imbalance between ego strength and oge strength is large a weak ego experiences oge-oppression as **depression**, and a strong ego experiences **elation** as it oppresses the oge; such oppression is an interaction with other empirical people, who represent the oge in the consciousness of the ego. A person with a very weak or non-existent oge is called a psychopath (by psychologists) or a sociopath (by sociologists); he or she has no moral sense and only behaves morally for learned prudential reasons, while behaving immorally — selfishly — when socially feasible. A person with a very weak ego, due to insufficient love when young, is schizophrenic, having delusions of persecution (because of being persecuted by the oge) and of grandeur (in order to bluff the oge); in dire circumstances such a person may go on a shooting spree, as a misguided pre-emptive strike, in self-defence against the oge; this is usually then quickly followed by suicide, which is a killing of the ego by an outraged oge.

Another form of internal conflict is neurotic conflict, of which the two main types are sexual (Sigmund Freud, (1856-1939)) and inferiority/superiority (Alfred Adler (1870-1937)). See my *Renascent Rationalism* for greater detail.

As in the previous chapter, we finish with a miscellany of explanations of psychological facts, with the same goal of demonstrating that the present theory of mind has reasonable scope and density of detail — two of the criteria of good explanation — and the miscellany has the same arrangement as in the previous chapter.

Anarchy is a desire to be free of the oge, a desire displaced on to the government, which is a representative of the oge.

Astrology. Those whose oge is sufficiently close to the ego, or strong enough, to make itself felt, often feel that there are influences beyond the blue sky which affect their fate. They attribute this to the effects of the planets, and are fascinated by astrology as a result. They are of course mistaken: neither the noumenal planets nor the empirical planets have any effect on their personal futures, as all rational people know; it is the oge that they sense, quite correctly, as affecting their fate — their oge being beyond the empirical blue sky, and their fate being the relative balance between ego and oge strengths and other content of the oge. Just how common this is, is shown by the number of astrological predictions available to the gullible in the press.

Divine right of kings. Monarchs used to believe that they had a divine right to rule because their coronation was a process of approval thereof by God. This seems reasonable, given that the oge is one of the meanings of the word god and a coronation is a public *rite of passage* in which the oge approves the monarch's right to rule.

Fashion. Fashionable people are following the dictates of their oge concerning the latest fashion. Leaders of fashion are ego-dominant and impose their creative ideas on the oges of their followers.

Feeling of being watched. Having this feeling when empirically alone is explained by the fact that the oge is watching. The feeling is particularly strong, naturally, when performing shameful actions.

Ghosts. Some people do genuinely experience ghosts; what they in fact experience is oge-memories causing empirical appearances in their empirical world — not actually surviving noumenal "spirits" of deceased people, which we can confidently say do not exist because they have little or no explanatory value.

Gossip is oge talking to oge.

Hypnotism is explained by the hypnotist representing the oge and putting the ego "to sleep" so that the oge may take control and make the individual behave according to the hypnotist's instructions. The fact that the hypnotist cannot make the individual behave immorally is due to the oge being the guardian of morals.

Malice provides an illusion of triumph over the oge, by diminishing the oge relative to the ego, as in practical jokes, bullying, and vandalism.

Politics. Oge-dominant people can be expected to be radical in politics: on behalf of their oge they favour the good of society, and particularly of the under-privileged, at the tax-payers' expense; ego-dominant people, on the other hand, selfishly want themselves to be over-privileged, with minimum taxes and maximum freedom, and so are politically reactionary. In each case extremists are passionate deniers of the other side, their passion helping to hide the weakness of their prejudicial position. Liberals, in the original sense of the word, prefer a path that is midway between these two extremes.

Primitive magic is attempts at dealing with the oge, all of which are displaced on to the empirical world. Sympathetic magic, as with pins in a voodoo doll, love-potions, and name magic — confusion between name and named — are attempts to control the oge by manipulating things or people in the empirical world. Belief in demons is belief in evil oge-persons, and some magic is attempts at control of demons.

Religion, in so far as it has congregations, is oge-worship by the ego. It is further discussed in Chapter 13.

Rites of passage. Baptisms, coming of age, graduations, weddings, and funerals are all, as rites of passage, necessarily public because they are social and so involve the oge. Public approval of these rites is oge approval.

Self-sacrifice is mis-named: it is sacrifice of the ego, by the oge, for the good of society. The ego is incapable of sacrificing itself because of the mind hekergy principle (180).

Sleep walking. When anyone is sleepwalking there is no puzzle as to how they can control their body while genuinely asleep: they do not control their body, the oge does. The same happens with sleep-driving.

Suicide is the killing of the ego, by the oge, for the good of society. It is necessarily preceded by depression (197).

Chapter 13. Gods.

Religious experience, although private and subjective, is so common that atheistic deniers of it have to be mistaken. We look here at its possible origins, taking into account the fact that if there are any actual Gods then they must be noumenal — simply because they are not empirical — and are so because of their explanatory value. There are in fact four features of the noumenal world which may be interpreted as God, without any lapse into authority, wishful thinking, fundamentalism, or superstition.

The first is the **oge-God,** or at least the top half of the oge, which is partly loving. It is beyond the blue sky, in heaven (196), and so transcendent to the empirical world, and its influence in the empirical world makes it also immanent. It is either a collection of gods (oge-people), as in the ancient Greek, Roman, and Nordic religions, and in the Hindu religion, or else it is unified into a single God. As an agent it is like a person and so a personal god, demanding worship and sometimes answering prayers if they are in its power to grant (most are not, but the oge will take credit for them if they are answered otherwise, as in prayers for rain). As an essentially social being, the oge-god demands that rituals be public, with congregations, in temple, synagogue, church, or mosque. Its strength lies partly in the size of its congregation — to a large extent it consists of memories of its congregation — so it tries to increase the size of its congregation by demanding missionary action of its adherents, and also by demanding their maximal reproduction. It may communicate with the ego by sending visions into the empirical world, such as an angel with a moral edict, or a voice from on high. It is most prominent in time of war, when society is in its greatest danger; then the upper half of the oge whips up patriotic fervour and the lower half of the oge — Satan — opportunistically takes partial charge, willing atrocities — war crimes — against the evil enemy. It is also prominent in promoting aid in catastrophes and crises. In oge-dominant people, for whom the oge is strong and close to the ego, this god may be felt as a presence. Ego-dominant people, on the other hand, are likely to be atheists, and even passionate deniers of any god. It is important to remember that there is one oge per noumenal mind, psychopaths perhaps excepted, so that a congregation is a group of noumenal people with similar oges; the similarities make the oge public to the group, who interpret these similarities as identity — through the identity error. Because oges are partly formed by introjection of other peoples's beliefs, to become oge beliefs, mostly those of parents and teachers, they are easy to change through the generations; this is the reason that there are so many schisms in religion (and politics). The oge is also a spirit: 'spirit' originally was breath, and identified with life, or agent — hence our word spiritual.

The second, and obvious, reference of the word God is the whole noumenal world: the **noumenal-God**. Since the noumenal world is the best of all possible worlds, this God is perfect. (But it is a mistake to call this perfection infinite: it is only maximal.) All causation, and upward and downward necessitation, is noumenal, so the noumenal world is all-powerful. Like the oge, it is both transcendent to, and immanent in, the empirical world; but it is not in heaven — heaven is in it. If the material is defined as anything concrete then this god is immaterial. But it is neither an agent nor a spirit, so is not a personal god; it does not demand worship and does not answer prayers. The noumenal world is a philosopher's God: it is the ground of all the being of empirical worlds. Given the causal theory of noumenal perception this God may fairly be described as the creator of noumenal minds, or souls, and hence of empirical worlds. It clearly was the God of Spinoza and of Einstein, the God that does not play dice.

The third meaning of the word God is $\top$, the top relation of the best of all possible worlds. Having necessary actuality it is *causa sui*, "self-caused"[23]. As such it is the first cause, of which all other causes and necessities are subsequent effects. Equivalently, it is the prime mover, the origin of all motion. It is the ground of all noumenal being, which includes all empirical being. Like the noumenal-God it is immaterial, neither an agent nor a spirit so not a personal God, and it neither demands worship nor answers prayers, and is not in heaven.

The fourth reference of the word God is to a third agent in the noumenal mind not so far discussed. This is the **psychohelios**, named after Plato's Form of the Good, which he likened to a sun in the mind because it both nourished and illuminated the other Forms (207). It is a structure of noumenal ideas, at the periphery of the ego, arranged with maximum hekergy and hence maximum rationality (168). It is created in the first place by an ego that is curious about reality: a philosopher, scientist, or any other like-minded person; but after it is big enough, it becomes an agent and grows by itself. Being of maximum hekergy it is similar to the noumenal world, which also is of maximum hekergy since it is the best of all possible worlds; so the psychohelios is true by similarity to the noumenal world, although it has vastly less density of detail (121). Hence as a God the psychohelios may be described as all-knowing: it is the god of truth. Truths that it has developed itself, as an agent, may be intuited (190) by the ego: as such they are most exhilarating discoveries for the ego. The psychohelios is also the god of the mystics, who claim that the consciousness of the ego may be transferred to this god, so that the ego becomes one with God: an ecstatic, ineffable, eternal, blissful, union with God. That is, a consciousness that is **supra-rational**, supra-linguistic, timeless, and of maximum possible value. Such a state is brought about by an ego that is mature enough both to form a

---

[23] This is not a monadic relation, but a misleading nominal meaning of original actuality.

psychohelios and to love (191) this god by giving unconditionally of itself — self-denial — to its beloved. This supra-rational state is what Plato meant by wisdom. This loving ultimately means the death of the ego and the rebirth of consciousness in the psychohelios; as such it is of maximum hekergy and so of maximum value, far more valuable than any empirical wealth, fame, or honour. It is not to be confused with the misunderstanding of the teaching of it, as the death of the empirical body followed by the resurrection of the empirical body, and ego, or soul, in Heaven, in the blissful, forever, company of the oge-god. (But note that there is some ground for this belief in resurrection: the deceased individual may be remembered by many as an oge-person, in the oge: for a long time, as with Abraham, Jesus and Mahomet, a short time as a family legend, or somewhere in between, as with Robin Hood. Or else as an evil oge-person, in Hell, in the lower oge, as with Stalin and Hitler. Such oge-people have a limited consciousness, in the sense that every noumenal idea reacts to other ideas, by L.A.L., but nothing as rich as the consciousness of the original ego of the person remembered as the oge-person.)

It is pertinent to remark that terms such as transcendent, immanent, heaven, hell, spirit, all-powerful, all-loving, all-knowing, perfection, ground of all being, creator, *causa sui*, self-caused, first cause, prime mover, mysticism, eternity (absence of passage of time), bliss, and ineffable (cannot be put into words) all have key importance in traditional theology.

It is, of course, a serious theological error to try to unite these four Gods into one: to do so is to commit the identity error, the denial of the principle that qualitative difference entails quantitative difference. The desire to so unite is an extension of the desire to unite the polytheistic, primitive, gods of the oge into a monotheistic god.

It is interesting that this theological analysis solves that most thorny of theological problems, the problem of evil. This is the problem that if God is all-loving, all-knowing, and all-powerful, and that evil exists, then being all-knowing He knows of the existence of evil, being all-loving He must want to abolish evil, and being all powerful He is able to abolish it. Yet evil exists. As a result there are then only three possibilities. He is all-loving but not all-knowing and/or not all powerful; He is all-knowing but not all powerful and/or not all loving; or else He is all-powerful but not all-loving and/or not all-knowing. Therefore there is no God which is at once all-loving, all-knowing, and all-powerful. But this problem is solved by the fact that the oge-God is all-loving (to its worshippers, its congregation, who are its chosen people) but neither all-knowing nor all powerful; the noumenal world is all-powerful but neither all-loving nor all-knowing; and the psychohelios is all-knowing (in low density of detail) but neither all-powerful nor all-loving.

And, finally in this chapter, what is evil? It is loss of hekergy. As perceived empirically, subjectively, and selectively by people, it is deplored. But in the noumenal world, which is the best of all possible worlds, hekergy is a maximum and conserved, so that every hekergy loss is balanced by an equal hekergy gain. The noumenal changes in hekergy which are imaged into empirical worlds as evil are just part of the flux of perfection.

Chapter 14. A Brief Historical Sketch.

This chapter selectively relates some of the history of philosophy to the present work.

Philosophy begins with puzzles in common sense and with attempts to explain them; puzzles such as the argument from illusion (172), the problem of secondary qualities (175), and the problem of identity and change. The problem of secondary qualities is the problem of whether secondary qualities — concrete qualities — are inside the head or outside. The problem of identity and change is due to the facts that qualitative difference entails quantitative difference (70) and that change is a qualitative difference over time (33). If X changes over time then the earlier X, X_1, is qualitatively different from the later, X_2, and so X_1 and X_2 are two — they cannot be identical, one and the same. So one thing cannot change with time: it either changes, and becomes two, or else it remains one and does not change. Because of this Heraclitus (c535-c475 BCE) claimed that nothing is permanent except the fact of change, while Parmenides (c515-c460 BCE) declared that only the One (identity) is, all change is illusion.

The view of Heraclitus is true of each person's empirical world.

The view of Parmenides is true of the noumenal world.

In saying that only the One is, Parmenides meant that only the One exists. He wrote that only Being is; and added that Being must be One because if it were many then it would be divided, which is impossible because Being cannot divide Being (similarities cannot divide similars) and non-Being cannot divide Being because non-Being is not, i.e., does not exist. Einstein's four-dimensional space-time is often remarked to be similar to Parmenides' One. It is likely, in my opinion, that Parmenides was one of the four western philosophers who achieved the supra-rational state of mind, the others being Socrates, Plotinus, and Spinoza.

A third pre-Socratic philosopher relevant to the present context is Pythagoras (c570-c495 BCE), famous for his theorem about right triangles and nearly as famous for his claim that "All is number". If, as is claimed in the present work, relations are the fundamental building blocks of the noumenal world and also the fundamental basis of number, *via* their characteristic property of adicity, then it is plausible that Pythagoras had an intuitive understanding of the nature of the noumenal world *via* his psychohelios (192).

Plato (428-348 BCE) spoke of *ideos* (ιδεοσ), which translates literally to our *ideas*; and because he meant something rather different from our *ideas*, his ιδεοσ is usually translated into our word *form*. His theory of forms was a theory of perfect knowledge, and from it we get our words *ideals* and *idealism*, referring to perfection and to goals of achieving it.[24] Plato only gave a few examples of forms in his writings, such as *bed, shuttle, justice, beauty* (in his dialogue *Symposium*), *temperance, courage*, and the *good*, but he did write (in his *Phaedo*) that there is a form of the true nature, or essence, of everything. The forms of bed and shuttle are ideal — perfect — in the sense of perfectly functioning artifacts, and the form of justice (in his *Republic*) is perfect harmony between the parts — either parts of a person, as in health, or parts of a society, as in a just society; these are all relations or properties of relations. The form of the good was particularly important to Plato; he likened it to the Sun, in that, as the Sun nourishes plants as well as illuminating them, so does the form of the good nourish and illuminate the other forms. If the form of the good is hekergy then in the best of all possible worlds the other forms have maximum hekergy and so are perfect. Plato did express doubts (in his dialogue *Parmenides*) about whether such things as hair, mud, and dirt had forms, but a possible elucidation of this is to say that the forms are top

---

[24] These words are also used technically in philosophy, meaning that the *ideal* is of ideas, and *idealism* is the philosophic doctrine that there is no matter, only mind, or ideas.

relations of wholes (26) or else properties of such relations — and hair, mud, and dirt are not wholes.

Plato was, in the opinion of many, including me, the world's greatest philosopher. His philosophy is important for at least two reasons that I want to emphasise: his two-world hypothesis, and his idea of wisdom.

His two-world hypothesis solves, in principle, two problems: the problem of identity and change (206) and the problem of illusion (171). These solutions are outlined in two famous passages in his *Republic*: the allegory of the cave, and the metaphor of the divided line.

In the allegory of the cave he asks you to imagine life as a prisoner in a cave, tied to a low wall behind which slaves move around carrying objects on their heads which throw shadows onto the cave wall in front of you, from the light of a fire behind them. The shadows of the slaves are not visible to you because they fall onto the wall behind you. If you could break free from your bonds, you could see the objects themselves rather than their shadows. Furthermore, if you could find your way out of the cave, you would see things by the light of the Sun; this latter is a metaphor for knowledge of the Forms, with the Sun as the Form of the Good. Just as the shadows in front of you, on the wall of the cave, are images of the objects on the heads of the slaves, so these slaves' objects are images of the Forms. And to know the Forms is to achieve wisdom. If Plato were writing today he might well liken the prisoners in the cave to couch-potatoes watching soap-operas all day on television: their world view is badly distorted because of seeing images of reality rather than reality itself. This "reality" is empirical reality (53), the reality of the empirical world that we each perceive around us, defined as all that is non-illusory in the empirical world — as opposed to the fictitious soap-operas. For Plato the empirical world is a moving image of eternity (*Timaeus*), while the world of the forms is an unchanging One. He thus provides a solution to the problem of identity and change: each person's empirical world is Heraclitean and the world of the forms — the noumenal world — is Parmenidean.

In the metaphor of the divided line he asks you to imagine a vertical line divided into two parts, in a certain ratio, with each part divided in the same ratio. The bottom-most part represents images of empirical objects, such as shadows and reflections (or, if he lived today, photographs, movies, radio, television, videos, and holograms), while the next part up represents the empirical objects, which are the originals of the images of which the bottom-most part consists. The upper portion of the whole line, divided in the same ratio, represents mathematics in its lower part, and knowledge of the Forms in its upper part. The point about "the same ratio" is that the whole line is analogous to the bottom portion, and also to the upper portion, and the upper portion is analogous to the lower portion. And the point about the lower portion is that shadows and reflections are images of empirical objects. So in the upper portion mathematics is an image of the world of Forms, and the whole lower portion is an image of this upper portion. Thus mathematics, our language of relations, is an image of the noumenal world, which consists of relations. And since mathematics is as rational as we can get, the only satisfactory designation of the topmost portion is to call it supra-rational (203). Wisdom, knowledge of the Forms, is supra-rational knowledge, perfect knowledge. The essential message of Plato's philosophy is that wisdom is possible. Note that φιλοσοφία, philo-sophia, or philosophy, means love of wisdom.

Plato was a student of Socrates (c470-399 BCE), who was condemned to death for corrupting the youth of Athens; the corruption is recorded historically as denial of the gods. I am of the opinion that the alleged corruption was S.B.S. (179), but that this could not be written down without dire consequences so "denial of the gods" was a safe and plausible substitute. In Leibniz (see below) S.B.S. is obvious in his *Monadology* if you know what to look for, but Leibniz did not dare draw attention to it. Bertrand Russell, living in a more tolerant age, did spell it out in *Human Knowledge*[25] but has been conspicuously ignored on this point. And I myself suffered professionally from advocating it. So S.B.S., in all its logical simplicity, is not a frequent topic in the history of ideas. However Socrates and Plato were both highly intelligent and, I believe, thoroughly understood it.

Aristotle (384-322 BCE), one of Plato's students and another very intelligent man, was much more inclined to common sense than Plato, and fell back on to naive realism. But he deserves mention in this sketch because of his invention of formal logic (see Appendix A), a subject-predicate logic which misled philosophers into a substance-attribute approach to metaphysics that has lasted to the present day. (Substances corresponded to logical subjects and attributes to predicates.) He assumed that every class has an intension, which he called the essence of the class; he thought more in terms of essences than of classes, so discounted contingent sets. He also wrote of form and matter, as well as substances and attributes. The matter of a table, for example, is pieces of wood, and the form of it is the shape of the table and the configuration of its parts; each piece of wood has matter and form, and so on down to prime matter; there had to be prime matter, he said, not itself composed of matter and form, because an infinite regress is impossible. The form of an object is the essence of the class to which it belongs. Form is as close as he got to dealing with relations; his logic, unlike mathematics, cannot in fact handle relations satisfactorily, and this led to many problems until Spinoza and Leibniz, each in his own way, and unsatisfactorily, solved the problem (see below). He also had a representational theory of perception, which supposedly resolved the question of whether we perceive real objects or images of them by claiming that we

---

[25] Bertrand Russell, "Human Knowledge, Its Scope and Limits", New York, 1948, Part 3.

perceive real objects *by means of* the images — a non-explanation unless the means can be spelled out, which they never have been. Aristotle's works were lost to the Christian world during the dark ages, but recovered from the more advanced Muslim civilisation in the medieval period and incorporated into Christian theology by St. Thomas Aquinas (1225-1274). Since then the Roman Church has accepted two bases of theology, one Platonic and one Aristotlean, with the Aristotlean one becoming more dominant.

Plotinus (c304/5-270 BCE) was the most important of a group of philosophers after Aristotle, called neo-Platonists. After Parmenides and Socrates, the only two other western philosophers for whom there is some evidence that they achieved wisdom, the supra-rational state of mind, are Plotinus and Spinoza. Probably as a consequence of this their writings are difficult to understand. Plotinus used the metaphor of emanation, from the usage that light and heat emanate from the Sun. From the One, Plotinus said, emanates nous (νουσ: intellect or mind), from which emanate the world soul and human souls, from which latter emanate matter. Emanation is from the more perfect to the less perfect, with the One being perfect — it is Plato's form of the Good — and matter being the least perfect, so that emanation includes a descent of quality. The key point of this doctrine is that a human soul may struggle up — ascend — to the One and so gain wisdom. This doctrine of emanation is mostly metaphor, and metaphor can be controversial because of multiple interpretations; but the interpretation offered here does seem to make more sense than others. It is: the One is the top relation of the noumenal world; and all other noumena, including nous, are descended (demerge) from the One. The One, clearly, is not the One of Parmenides, since it is not the totality of Being; but as T it is the single relation at the top of the noumenal world which, being single, cannot form an emergence-configuration and so cannot emerge any uppers; so it makes the noumenal world complete. All other noumena are at lower levels than T, and

their actuality, or Being, demerges from T because ⊡T — T
has intrinsic actuality — so these lower noumena may well be
described as descending from T. Then nous, or intellect, is
noumenal mind; the world soul is the psychohelios; human
souls are noumenal egos; and matter is the stuff of empirical
worlds, which are partial images of the noumenal world,
within the noumenal ego — this stuff, or matter, being
secondary qualities, which are everything concrete. And the
consciousness of the noumenal ego may become one with the
consciousness of the psychohelios, in a state of supra-rational
knowledge of the noumenal world, or wisdom.

Descartes (1596-1650) is famous for making a new
start in philosophy, after medieval philosophy had dwindled
down into a triviality comparable to the triviality of twentieth
century philosophy. He sought absolute certainty as a starting
point for philosophy and found it by doubting, as a philosophic
exercise, everything that he possibly could, which resulted in
one indubitability: his *cogito ergo sum*, "I am conscious[26]
therefore I exist", which he declared to be the one absolute
certainty; because if he did not exist he could not be
conscious, hence his own existence was absolutely certain.
From this he argued that other certainties could be deduced,
such as the existence of God, deduced using the ontological
argument. Descartes was also a mathematician — he invented
co-ordinate geometry, thereby uniting algebra and geometry
into one subject — and he suggested that metaphysics should
be mathematical rather than logical. However his own
metaphysics was logical: he said that there are two substances,
which he called thought and extension but which today are
better known as mind and body. Two points relevant to the
present discussion arise from this.

---

[26] To be literal, *cogito* means 'I think', not 'I am conscious'; but
in Descartes' day there was no Latin word for consciousness, and it is clear
from the context of Descartes' writings that he meant consciousness when
he wrote *cogito ergo sum*.

One point was the rise of modern science, with the work of Copernicus (1473-1543) and Galileo (1564-1642). Copernicus, like Aristarchus (c. 310 – c. 230 BCE) argued that the Earth is circling the Sun, as opposed to the traditional view that the Earth is the fixed centre of the Universe; and Galileo agreed, and also argued that science should not rely on Aristotle and other ancient authorities but should be done by reading the book of nature, a book written in the language of mathematics. (For Galileo science was exclusively empirical, while the search for underlying causes should be left to the philosophers.) Descartes was enthusiastic about science, unlike the Roman church, which was too rigid to change from its reliance on Aristotle. A case can be made that Descartes, a devout Catholic, deplored the looming conflict between science and Church, early manifested in the Church's condemnation of Galileo, and tried to defuse it by dividing the world into two non-interacting arenas, or substances: mind, for the Church, and matter, for science. He called these thought (or consciousness) and extension. However this led to difficulties for Descartes: mind and matter do in fact interact — for example, mind acts on matter when it wills its body to move, and matter acts on mind when drinking wine. Descartes later postulated a third substance, human nature, which served as a mediator between mind and matter; but this does not work because interaction is a relation, not a substance. To this day this difficulty is known as the mind-body problem. Descartes also distinguished two kinds of knowing: confused images, and clear and distinct ideas. Confused images are concrete — he gave the examples of man, horse, and dog — and clear and distinct ideas are abstract, of which mathematical ideas are examples. Descartes is important in the present work because of the significance of his *cogito* in Theorem 9.1 (148).

Spinoza (1632-1677), the greatest Dutch philosopher and the greatest Jewish philosopher (in my opinion), was the fourth philosopher in the history of western philosophy for whom there is evidence of having achieved supra-rationality.

He much admired Descartes but recognised the inadequacy of his two substance approach. Instead he said that there is only one substance, which he called God. God, or the noumenal world, has an infinity of attributes, of which we can know only two: thought and extension. A possible explanation of Spinoza's God having an infinity of attributes is to say that thought and extension are emergent relations in our noumenal minds, but that there are an unknown number of other novel emergent relations in the noumenal world. Being unknown, this number is infinite. Or, more kindly, it is the maximum possible number of novel emergents, which is perfection. Spinoza said that we have three kinds of knowing: *imaginatio* (imagination), as in man, horse, dog; *ratio* (reason) or clear and distinct ideas, as in mathematics; and *scientia intuitiva* (literally, intuitive science) which is supra-rational — although he did not use this word. He also distinguished between the free and the unfree: we are free when the cause of our behaviour is internal, and unfree when it is external. To behave freely is action and to behave unfreely is passion; in action we are an agent, in passion we are a patient. (This last usage survives today only in medicine.) We can achieve *scientia intuitiva* by control of the passions to the point of becoming wholly free. In this state we have a perfect knowledge of God, which must be Platonic wisdom, or the supra-rational state.

Gottfried Wilhelm Leibniz (1646-1716) was (in my opinion) the greatest German philosopher. His system can be derived from one axiom: namely, all truth is analytic. From this we can deduce the following, using the traditional logic that Leibniz used.

1.   All true propositions, being analytic, have their predicates contained in the subject and so must be the universal affirmative categorical "All S are P" propositions of traditional logic (see Appendix A).

2.   Hence the principle that Leibniz called *The Principle of Sufficient Reason*. Because the existence of a whole is a sufficient condition for the existence of its parts, the existence of a subject is a sufficient condition, or reason, for the existence of each of its predicates. Consequently this Principle means that for every predicate there must be a subject, which is the sufficient reason for that predicate; and to state this sufficient reason is to explain that predicate.

3.   All factual truth must be analytic — that is, all statements about what exists must be analytic. So whatever exists must be a substance having attributes, such that an adequate idea of the substance contains an adequate idea of each attribute of this substance. These adequate ideas are subject and predicates respectively. Leibniz gave as an example of this the claim that an adequate idea of Julius Caesar contains the idea of his crossing the Rubicon, so that the historical fact of Caesar's crossing the Rubicon can in principle be deduced from the adequate idea of Caesar. Such an adequate idea is of infinite magnitude however (see 5 below) and hence not possible within human consciousness.

4.   The total number of substances that exist in reality is either a minimum or a maximum, or some number in between. Any number in between would be arbitrary — there would be no sufficient reason for it. The minimum number of substances is zero, and it is false that zero substances exist, since I exist — my own existence is indubitable: *cogito ergo sum*. Consequently a maximum number of substances exists, and this maximum number is infinity. Each of the infinity of substances Leibniz called a **monad**. A monad is a soul and so conscious.

5.   An analogous argument applies to the number of attributes of each substance. This number is neither arbitrary

nor zero, hence infinite. It follows that an adequate idea of
each substance contains an infinity of predicates.

Another way of putting items 4 and 5 is to say that a
world consisting of an infinity of substances, each with an
infinity of attributes, is the best of all possible worlds. Best in
the sense that any alternative possible world would have to
contain fewer attributes. Hence the sufficient reason for this
world's existence is that it is the best of all possible worlds.
This dictum, which is perhaps the best-known part of Leibniz,
should not be confused with Voltaire's misunderstanding of it
in *Candide*, where it becomes a superficial justification of
laissez-faire conservatism. Voltaire was a common sense
naive realist, who did not understand that Leibniz's best of all
possible worlds referred, not to an empirical world of a
monad, but to the noumenal world. Empirical worlds are poor
images of small portions of the noumenal world and so do not
reveal its perfection. Leibniz gave the analogy of a huge and
beautiful painting, which is hidden by a cloth that has a small
hole in it; what you see through the hole gives you little idea
of the whole painting or of its beauty.

6.   Because all true propositions are categorical and
analytic there are no true propositions about relations, which is
to say that no relations exist. All relations are *entia rationes*,
said Leibniz — things of the mind, hence illusory.

7.   Consequently there are no intrinsic relations in a
monad: that is, every monad is indivisible — it has no parts, it
is a simple substance. (An attribute of a substance is not a part
of it, for substance-attribute metaphysicians: attributes
"inhere" in substances, and inhering is not a relation. In the
present context we have it that relational intrinsic properties
inhere in relations, but this does not mean that relations *are*
substances, because relations have terms, while substances do
not.)

8.   Also, there are no extrinsic relations between monads — for example, no spatial or temporal relations, no causation and no similarities or dissimilarities. Leibniz described this metaphorically by saying that every monad is *windowless*.

9. So the infinity of monads occupies no space at all, and is timeless — that is, eternal.

10.   And there is no causation between one monad and another, each is wholly independent of every other. Because there are no causes all explanations are by means of sufficient reasons.

11.   And no two monads are alike, since there is no similarity between any of them. This is Leibniz's famous *Principle of the Identity of Indiscernibles*: no two things differ numerically alone. If they are two, it must be because of a qualitative difference — a difference of attributes. It should be noted here that for Leibniz similarity would be exact similarity; anything else is dissimilarity, and dissimilarity is not a relation, since it is merely the absence of similarity. Consequently every monad has some characteristic, qualitative difference from every other. This characteristic difference Leibniz called its *viewpoint*.

12.   If monads were to be ordered according to the closeness of their viewpoints, they would seem to be in a spatio-temporal-causal ordering. This ordering Leibniz called the *pre-established harmony*.

13.   Because both the noumenal world and each monad are, as it were, of maximum richness, each monad is in this respect a copy of the world; it *mirrors* the world, as Leibniz said. Consequently among its attributes each monad contains a copy of every other monad plus a copy of the pre-established harmony; yet at the same time each monad differs from every

other. So each monad contains a spatio-temporal-causal copy of the noumenal world, a copy which is unique in its viewpoint. Or, to put it another way, the pre-established harmony is a spatio-temporal-causal assemblage of viewpoints, which, when mirrored, is mirrored at and from each of these viewpoints. As such, it is, in each mirroring, the empirical world in the consciousness of the monad at that viewpoint.

14.   Within each mirroring there are things — empirical objects. Things are mirrorings of assemblages of monads of closely related viewpoints. Such assemblages Lcibniz called compound substances. They are compound as compared with simple substances, or monads.

15.   Finally, among monads are human souls. These perceive, each in its own world, its own human body. This empirical body is a mirroring of a compound substance, the soul's noumenal body. Each such compound substance consists of an infinity of monads, so it follows that each human being has an infinity of souls — a strange doctrine that Leibniz disguised in his published writings by the supposition that one of these monads is dominant over all the others. This dominant monad was then the one soul required by Christian teaching.

Leibniz was marvellously consistent, and his consistency led him to conclusions such as: the noumenal world exists necessarily because it is the best of all possibles, and there is one empirical world per consciousness. So why then is his system unacceptable in the present context? There is just one thing wrong with it: its rationality is Aristotelian logic rather than mathematics, so that his noumenal world consists of substances having attributes rather than of relations and structures. And, ironically, his system is so consistent that it is almost self-healing. Leibniz did not really reduce all relations to *entia rationes*; instead he put all of them into the pre-established harmony. The pre-established harmony is an infinite-adic relation which relates every one of the infinite attributes of every one of the infinite substances, or monads, to every other one of the infinite attributes of every other one of the other infinite substances. So analysis of the pre-established harmony should lead to a noumenal world of relations, described by mathematics. Leibniz, the mathematician who invented calculus, determinants, and binary numbers, had too much respect for Aristotle and his logic.

David Hume (1711-1776) was a Scottish philosopher in the empiricist tradition of Francis Bacon (1561-1626) and John Locke (1632-1704). His thought may be justly described as being derived from the exact opposite of Leibniz' axiom; namely, all truth is synthetic, as opposed to Leibniz' axiom that all truth is analytic. Hume phrased his axiom as: no idea without an antecedent impression. By impression he meant, following Locke, a perception. In other words, there are no innate ideas, no inherited ideas; the mind at birth is like a blank wax tablet, as Locke put it, and perceptions are impressions on that tablet. So everything we know comes through the senses, and all speculation about anything that cannot be perceived is futile. So 'tis vain to speculate concerning the secret springs whereby bread nourishes man. Hume was temperamentally a sceptic and this approach suited his scepticism well. But he was inconsistent in his scepticism: he allowed that he had an idea of the Self, even though he had no antecedent impression of it, and he allowed the common sense metaphysical (hence speculative) beliefs in the continued existence of empirical objects when unperceived and in the existence of minds in other empirical people. If he had been more consistent on these he would have been a solipsist (see Appendix C); he was in fact on the slippery slope to solipsism, with nothing to prevent his sliding to the bottom except dogmas of common sense. His importance in the history of philosophy was his influence on Kant, as well as on subsequent empiricism.

Immanuel Kant (1724-1804) was a German philosopher of Scottish descent who is widely regarded as being the most influential philosopher of modern times. He was a Leibnizian scholar until, as he put it, he was awoken from his dogmatic slumbers by David Hume, and then he sought to reconcile Hume and Leibniz. He was Leibnizian in denying the reality of relations but not in regard to Leibniz' ideas on perception; and he was Humean in denying the possibility of knowledge of the noumenal, but not in denying the innateness of what he called the *a priori* synthetic. Knowledge, he claimed, may be characterised in four ways: it may be *a priori* or *a posteriori*, and it may be synthetic or analytic. It is *a priori* if it can be known prior to perception, and it is *a posterori* if it can be known only after perception; and it is analytic if it is known through analysis and synthetic if known through synthesis. He claimed that empirical knowledge is *a posteriori* synthetic, and mathematics and logic are *a priori* analytic. But, he wrote, there is also *a priori* synthetic knowledge, which is innate, or genetic as we might say today, or hard-wired in our brains, as computer people might say. It is the way our minds have to work, whether we like it or not. The way Kant put it is that space is our *a priori* synthetic form of outer intuition, time is our *a priori* synthetic form of inner intuition, and the categories are our *a priori* synthetic form of understanding. By outer intuition he meant what we call perception, by inner intuition he meant introspection, and the categories are a somewhat artificial list of ways of thinking. In other words, we cannot perceive the empirical world except spatially, we cannot introspect except temporally, and we cannot think without the categories. Space, time, and the categories are all relational, and because they are *a priori synthetic* they are thereby *entia rationes*, as Leibniz put it, things of the mind. Kant argued that inner and outer intuition and the categories are *a priori* synthetic because it is a fact that *everyone* perceives, introspects, and thinks in these ways.

Kant can be faulted in at least three ways. The first is that his claim to the *a priori* synthetic is based on universality, which is not necessity hence it allows other possibilities, such as some relations being empirically real entities. That is, universality is extensional necessity, as opposed to relational necessity, which is singular possibility; and the claim that *everyone* is conditioned by the *a priori* synthetic is only extensional necessity, not relational necessity. The second fault is that the *a priori* synthetic requires that each consciousness has its own, numerically distinct empirical world, which is the Leibnizian view. In Kant these empirical worlds must be numerically distinct because each empirical world is generated *a priori* synthetically by a numerically distinct mind, but Kant simply assumed the common sense view on this, uncritically: there is only one empirical world, containing empirical people, each of whom is conscious of their locality in it, from their own viewpoint, their own here and now. Third, his Humean denial of any possible knowledge of the noumenal put him, like Hume, on the slippery slope to solipsism with nothing but dogma to stop him sliding to the bottom. That is, the existence of conscious minds in empirical people other than oneself cannot be known by oneself; so not only can Kant not possibly know that the *a priori* synthetic is universal, he cannot know of the existence of anything outside of his own *a priori* synthetically controlled consciousness.

Kant was followed by Georg Wilhelm Friedrich Hegel (1770-1831), famous for his idea of dialectic, consisting of tripartite steps of thesis, antithesis, and synthesis. This was a matter of "sublation" (*aufhebung*) of opposites: a process of resolving the opposites while preserving them and elevating them. This led to a series of sublations, beginning with the idea (thesis) of *being*, then of its opposite (antithesis), *nothingness*, then of their synthesis, *becoming* (synthesis). Hegel is noteworthy in the present context because this process of repeated dialectic ended with the sublation of the finite self into the infinite absolute — a result best described as supra-rational. Hegel was also significant in introducing what is now known as a Hegelian change: a sufficient change in quantity may produce a change in quality — the latter being an emergence.

The most important philosopher in the twentieth century) was, in my opinion, Bertrand Russell (1872-1970). He is important in a number of ways but only one will be mentioned here: his analysis of perception. His early analysis[27] (1918) required a six-dimensional space: using the adjectives *physical* and *perceptual*, for the present noumenal and empirical, he said that at each point in three-dimensional physical space there is a three-dimensional perceptual space, so that to locate something in perceptual space required six numbers: three in perceptual space and three more to locate that perceptual space in physical space. This is all that he meant by six-dimensional space. However, most philosophers know little mathematics, and, along with other non-mathematicians, they found the idea of six-dimensional space both incomprehensible and intimidating; so this idea did not win many adherents. Russell was thinking in this way because the fact of illusion requires a duality of real object and illusory image of it, which is to say *physical* object and *perceptual* image. And since physical objects occur in physical space and perceptual images occur in perceptual space, perceptual spaces are both images *of* physical space and *in* physical space. His later analysis of perception[28] (1948) omitted the need for a six-dimensional space by making perceptual spaces portions of physical space — an improvement over his earlier analysis since a perceptual space at every point in physical space

---

[27] Bertrand Russell, *Mysticism and Logic*, London, George Allen and Unwin, 1918.

[28] *Human Knowledge*, London, George Allen and Unwin, 1948.

multiplies entities beyond necessity. In more detail, physical space contains physical human beings whose brains each contain a perceptual space which is an image of their local physical space and which contains images of physical objects — including an image of their own physical body. Thus Russell's physical space is the noumenal world and his perceptual spaces are empirical worlds.
As Russell writes[29]:

> "One of the difficulties which has led to confusion was failure to distinguish between perceptual and physical space. Perceptual space consists of perceptual relations between parts of percepts, whereas physical space consists of inferred relations between inferred physical things. What I see may be outside my percept of my body, but not outside of my body as physical thing."

Compare this with "Russell's joke" on page 192.
Another twentieth philosopher, important historically for his misdirection, was G. E. Moore (1873-1924). He argued that the over-simplifications of common sense are in fact truisms, obviously true, and so the proper basis of philosophy. This, with support from Wittgenstein, led to wide-spread acceptance of naive realism (more usually called by its advocates 'common-sense realism', or simply 'realism') in English language philosophy. Along with phenomenology and existentialism on the continent, which also take naive realism for granted, this erroneous assumption is thus almost universal in present western philosophy, to its shame.

---

[29] *Ibid*, pp. 224-5; Routledge edition, Oxford 2009, p.185.

Appendix A: Traditional Logic.

Traditional logic, which originated with Aristotle, is based on **categorical propositions**, of which there are four kinds: 'All S are P', 'No S are P', 'Some S are P', and 'Some S are not P', where 'S' stands for subject and 'P' for predicate; each of these latter is called a **term** and refers to a class of entities, so that this logic is also called a logic of classes — from which we get the word classification. Examples are 'All men are mortal', 'No cats are dogs', 'Some flowers are annuals', and 'Some primates are not apes'. Medieval logicians, who believed that all teaching should be by rote, developed mnemonics for much of traditional logic. The categorical propositions were symbolised by SaP, SeP, SiP, and SoP, in the same order as given above. The first two categorical propositions are called universal propositions, and the last two are called particular propositions; and the first and third are called affirmative propositions, and the second and last are called negative propositions. Whether the proposition is universal or particular is called the quantity of it, and whether it is affirmative or negative is called its quality. If universal propositions are ordered with first priority, affirmative propositions with second priority, and particular propositions with third, then the vowels in the symbols SaP, SeP, SiP, and SoP are in alphabetical order. Two special cases of a universal affirmative proposition are singular propositions and identical propositions: a singular proposition has a one-membered class as its subject (e.g. Socrates is mortal) and an identical proposition has the identical class for subject and predicate (e.g. All people are human). A term is said to be **distributed** in a categorical proposition if, in that proposition it refers to the entire class named, and otherwise it is undistributed. For example, in 'All men are mortal' the reference is to all men but not to all mortals, so the subject is distributed and the predicate is undistributed. This is indicated when necessary by dSaPu, and this is true of all A-propositions, other than identical propositions which are dSaPd. The others are: dSePd, uSiPu, and uSoPd.

There are three kinds of valid **immediate inference** in traditional logic (so called because the inference goes immediately from premise to conclusion): **obversion**, **conversion**, and the **square of opposition**. In obversion the predicate is negated — P becomes non-P, written with a bar over it — and the quality is changed; a negated predicate is, of course, the complement of that predicate. In conversion S and P are interchanged and the quality is unchanged, provided only that in the process an undistributed term does not become distributed — which means that an A and I propositions convert (*per accidens*) to an I, and that an O proposition does not convert. Alternate obversions and conversions produce chains of immediate inferences, as in Fig. A1, where a small o stands for obversion and a small c for conversion.

The square of opposition simply shows the truth or falsity of other categorical propositions —

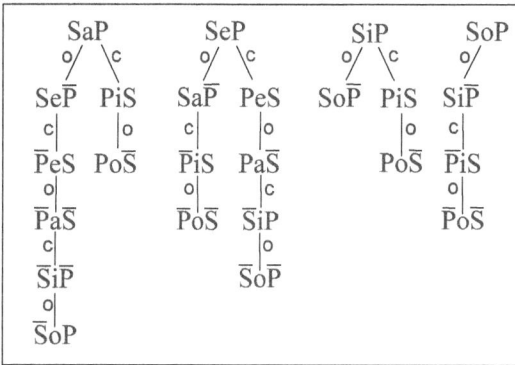

Fig. A1.

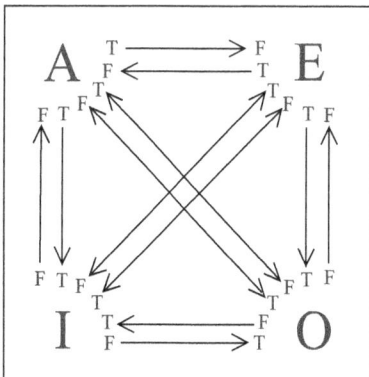

Fig. A2.

225

A, E, I, or O — with the same subject and predicate, as in Fig. A2.

Arguments with multiple propositions in traditional logic are called **syllogisms**, of which **categorical syllogisms** are the most important. (Others, not discussed here, are hypothetical and disjunctive syllogisms, dilemmas, and sorites.) A categorical syllogism, hereafter simply called a syllogism, consists of two categorical propositions as premises and one as a conclusion. Aristotle's famous example is 'All men are mortal and Socrates is a man, therefore Socrates is mortal'. The parts of a syllogism are all named: the predicate of the conclusion is called the **major term** (P) and the subject of the conclusion is the **minor term** (S); the reason for these names is that if S and P are thought of as sets, then $S \subset P$ and P is larger than S. The premise which contains the major term is the **major premise**, and is written first, and the other premise is the **minor premise**, which contains the minor term and is written second; and the third term is called the **middle term** (M), and occurs in each premise. Thus a syllogism might be MaP, SaM, therefore SaP. Complete classification of syllogisms is achieved with two further definitions. The **mood** of a syllogism is given by listing the vowels in the major premise, minor premise, and conclusion, in that order: thus the mood of MaP, SeM, therefore SoP is AEO. And the **figure** is a number from 1 to 4, given by the arrangement of the middle terms, as in Fig. A3. A mnemonic for this is that the lines through the M's in Fig. A3 look like a shirt collar.

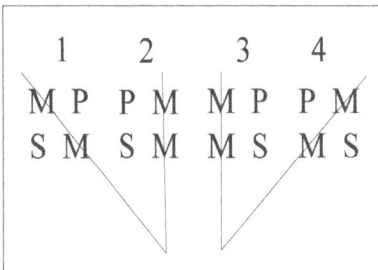

1	2	3	4
M P	P M	M P	P M
S M	S M	M S	M S

Fig. A3.

There are four rules that must be met for a syllogism to be valid. They are:

1. The middle term must be distributed at least once; if it is not the **fallacy of undistributed middle** occurs. For

example: All men are human, all women are human, consequently all women are men.

2. If the major term is distributed in the conclusion then it must be distributed in the premise; otherwise the **fallacy of illicit major** occurs. E.g: Some humans are women, no women are men, hence no men are human.

3. If the minor term is distributed in the conclusion then it must be distributed in the premise; otherwise the **fallacy of illicit minor** occurs. As in: All warriors are soldiers, some corporals are warriors, therefore all corporals are soldiers.

4. The number of negative conclusions (0 or 1) must equal the number of negative premises; otherwise the fallacy of the number of negatives occurs. For example: No flowers are animals, no animals are immortals, hence no immortals are flowers.

An example of a syllogism in which all four fallacies occur is: Some wine is red, some flowers are red, therefore no flowers are wine.

There are 256 possible syllogisms, of which 16 (or 18) are valid. The medieval logicians created a wonderful mnemonic for those eighteen: "Barbara, Celarent, Darii, Ferioque *prioris*; Cesare, Camestres, Festino, Baroco *secundae*; *Tertia* Darapti, Disamis, Datisi, Felapton, Bocardo, Ferison habet. *Quarta insuper addit* Bramantip, Camenes, Dimaris, Fesapo, Fresison." The mood of each syllogism is given by the vowels in its name — Barbara has mood AAA — and the figure is given by the Latin *prioris*, *secundae*, *tertia*, and *quarta*. (Two of these: Darapti, AAI-3, and Bramantip, AAI-4, are not really syllogisms in their own right because they are merely weakened versions of Barbara with a different figure; if they are excluded from the count, the number of valid syllogisms decreases from 18 to 16.)

The four syllogisms in the first figure — Barbara, Celarent, Darii, and Ferioque — may be regarded as axioms, and all the others may be deduced from them, according to various letters in their names. To begin with, the initial letter — B, C, D, or F — of a syllogism shows which first figure syllogism it may be derived from: thus Darapti, Disamis, Datisi and Dimaris are derivable from Darii. Other letters show how the derivation is achieved: an *s* after a vowel means that that vowel is simply converted, a *p* means that the vowel is converted *per accidens* — from an A to an I, or an I to an I — an *m* means that the premises must be transposed, and an *s* at the end means that the conclusion must be converted. Thus Cesare and Camestres are derived from Celarent by converting the major premise in Cesare and the conclusion in Camestres, and in Camestres the premises are transposed. Bramantip derives from Barbara, with the third A converting to I and the premises being transposed. Baroco and Bocardo are more complicated: they can only be derived from Barbara by *reductio*, by assuming that the conclusion does not follow from the premises and then deriving a contradiction.

The modern approach to syllogisms is Venn diagrams, in which S and P, or S, P, and M, are represented by intersecting circles, as shown in Fig. A4. Here the first four Venns give the four categorical

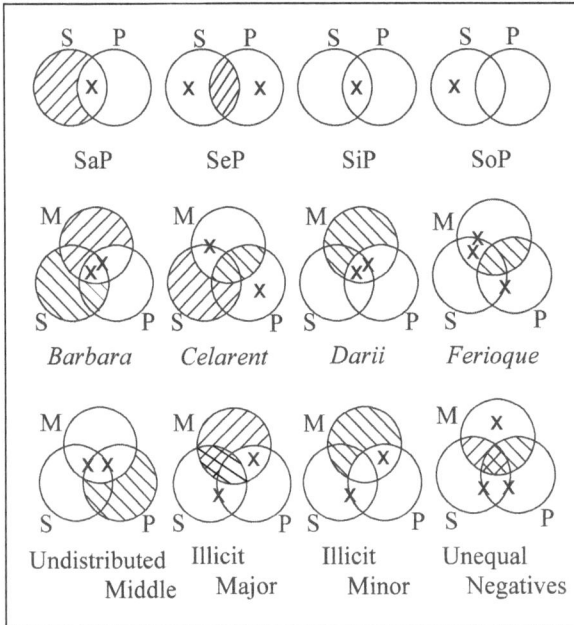

Fig. A4.

propositions, the next four give the valid syllogisms of the first figure, and the last four give examples of fallacies. In all of these shading of an area shows that area to be empty, and an x in an area shows it to be not empty. If it has neither shading nor x then it is unknown, and if it is unclear on which side of a line an x should be, it goes on the line. Traditional logic, unlike modern quantificational logic, has **existential presupposition** (56): it is presupposed that in a universal proposition it is true that least one S and at least one P exist. In showing a syllogism, both premises are entered into the Venn, including existential presupposition, and if the conclusion appears then the syllogism is valid, and otherwise it is fallacious. Thus the four Venn diagrams for the fallacies in Fig. A4 show no conclusions.

The metaphysical application of traditional logic to the noumenal world, in which subjects become substances and predicates become attributes of substances, fails because this logic cannot handle relations satisfactorily. Consider relations of direction: to say that Montreal is north-east of Toronto is to have two subjects/substances, Montreal and Toronto, and a relation, N-E, between them. We could say that Montreal has the attribute of being N-E of Toronto, but this turns a subject, Toronto, into an attribute. Equally we could say that Toronto has the attribute of having Montreal to the N-E, but this has the same objection. Or we could say that a substance, Canada, has the attribute of having a city, Montreal, to the N-E of a city, Toronto. None of these are satisfactory, but the third is necessarily repetitious: Canada becomes an attribute of planet Earth, Earth becomes an attribute of the solar system, and so on — until we finish up with the ultimate substance, the entire noumenal world. This was Spinoza's solution (213). The other alternative is to say that there are no noumenal relations, only unrelated substances; this was Leibniz's solution (214). A blend of the two is the many-worlds explanation (160) of cosmic coincidences: each such world is a substance and there are no relations between such worlds — they are windowless, as Leibniz would say (217). And the most satisfactory solution is to say that a logic of classes, or sets, is inapplicable to the noumenal world: there are neither noumenal substances nor noumenal attributes, because metaphysics and theoretical science must be mathematical and because mathematics has much more explanatory value than logic.

Appendix B: Truth-functional Logic.

Truth-functional logic begins with statements, which are grammatically correct sentences which are either true or false[30].

A simple statement, represented by lower case letters such as p, q, r, may be compounded with another into a compound statement, using five connectives:

'Not-p' symbolised by ~p.
'p and q' symbolised by p∧q.
'p or q' symbolised by p∨q.
'If p then q' symbolised by p→q.
'p if and only if q' symbolised by p≡q.

(~p is not a compound of two statements, but is still called a compound.)

These symbols are **truth-functions**, whose domain is the truth values of p and q and whose range is the truth values given in Table B1:

p	q	~p	p∧q	p∨q	p→q	p≡q
T	T	F	T	T	T	T
T	F	F	F	T	F	F
F	T	T	F	T	T	F
F	F	T	F	F	T	T

Table B1.

Thus ~p, called a negation, is false if p is true, and true if p is false; p∧q, called a conjunction (with p and q called conjuncts), is read as 'p and q' and is true if both p and q are true and is otherwise false; p∨q, called a disjunction (with p

---

[30] Some sentences, such as questions and commands, may be grammatically correct but neither true nor false.

and q called disjuncts), is read as 'p or q', is false if p and q are both false, and is otherwise true; p→q is called a material implication (with p called the antecedent and q called the consequent), is read as 'if p then q' or 'p implies q', is false if p is true and q is false, and is otherwise true; and p≡q is called an equivalence, is read as 'p if and only if q', and is true only if p and q have the same truth value.

Compound statements may be compounded into still larger statements, using parentheses (), brackets [], or braces {} as punctuation.

An inference is represented by material implication, →. It is valid if it is a tautology (always true) and is otherwise invalid.

An equivalence is represented by equivalence, ≡. It is valid if both components are true, or both false, and invalid if one is true and one is false.

Some examples are:

> If it is hot (h) then you will wear less (w): h→w.
> If it is hot (h) and raining (r) then you will not go out
> (~g): (h∧r)→~g.
> If it is hot (h) and raining (r), and windy (w), then you
> will read (p) or nap (q): [(h∧r)∧w]→(p∨q).
> It is an equilateral triangle (l) if and only it is an
> equiangular triangle (a): l≡a.

A **logical proof** consists of one or more given statements, called premises, as well as a conclusion, and intermediate inferences establishing the truth of the conclusion, given the truth of the premises. The proof is set out formally in a series of lines, each consisting of three parts: a line number, for later cross-reference, a logical statement, and a logical justification for that statement. Logical justifications are premises, previously proven argument forms, previously proven equivalences, previously proven theorems, or else axioms, definitions or tautologies. Statements are defined as true unless preceded by ~; that is, the absence of '~' signifies truth. Argument forms and equivalences are formulations well known to be valid, or may be theorems that are proved in a formal logical calculus (66). Some examples are:

Modus ponens: [(p→q)∧p]→q.
Modus tollens: [(p→q)∧~q]→~p.
Reductio: [p→(q∧~q)]→~p.
Simplification: (p∧q)→p.
Equivalence: (p≡q)≡[(p→q)∧(q→p)].
Double negation: p≡~~p.
DeMorgan: (p∧q)≡~(~p∨~q).

As well, any tautology may be introduced as a line, as may a previously given axiom or previously specified definition. Any earlier line or lines may be invoked in a later line (with two exceptions, given in the next paragraph). For example:

1. p→q   Premise.
2. p   Premise.
3. q   1, 2, M.P.

where M.P. stands for *modus ponens*[31].

Two special forms of deduction are conditional proof and indirect proof. Each of these begins with an assumption, specified as such: "Assumption C.P." or "Assumption I.P." The line on which one of these is introduced is indented, as are all subsequent lines, up to and including the last line of the special proof. This is to remind the reader that these lines are dependent upon the assumption and so may not be referred to

---

[31] *Modus ponens* is short for *Modus ponendo ponens*, which translates to 'The method of by affirming I affirm'. Similarly, *Modus tollens* is short for *Modus tollendo tollens*, meaning 'The method of by denying I deny'.

once the special proof is ended. In conditional proof a statement, p, say, is assumed and another statement deduced, q, say; the proof then ends with the (unindented) line p→q; in other words, by assuming p one gets q so one can say p→q, "If p then q". This last line is justified by invoking the first and last indented line numbers and "C.P." In indirect proof the object is to prove the assumption false by deriving a contradiction — a *reductio ad absurdum* in other words; if p is the assumption and q∧~q ends the indirect proof then the next (unindented) line is ~p and is justified by invoking the indented line numbers and "I.P."

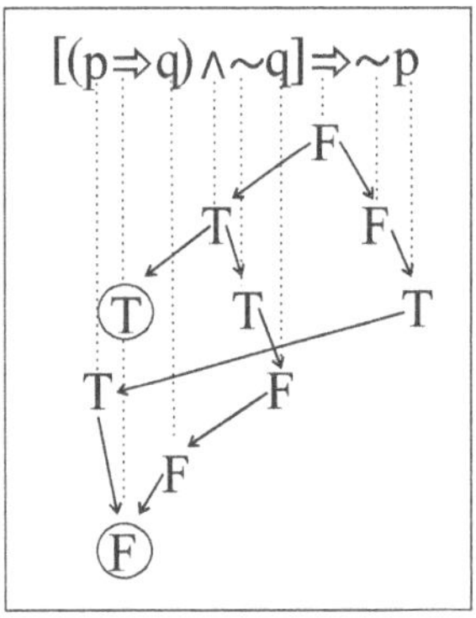

$$[(p \Rightarrow q) \wedge \sim q] \Rightarrow \sim p$$

Fig. B1.

That an argument form is a tautology is easily proved in truth-functional logic, as illustrated in Fig. B1, in which the validity of *modus tollens* is proved by *reductio*, using a tree of truth-functional consequences. That is, assume that the argument form is not a tautology, by putting an F under the primary connective, ⇒; this requires the secondary connectives to be the antecedent, ∧, which has to be true, and the consequent, ~, which has to be false, which in turn requires p⇒q to be both true and false, as indicated by the circled T and F. This contradiction proves that the primary connective cannot be false, so is a tautology.

The use of the three-part lines consisting of number, statement, and justification, and the use of logical symbols, are the most valuable innovative features of modern logic. The use of quantifiers and variables is another innovative feature of modern logic, discussed next.

Quantificational logic, or the monadic predicate calculus, uses two quantifiers: the universal quantifier, symbolised by (x) or (∀x) and reading " For every x" or "For all x"; and the existential quantifier, symbolised by (∃x) and reading "There exists an x" or "For at least one x"; x in these expressions is a variable. These quantifiers presuppose the existence of a universe of discourse, U, which is the totality of possible grammatical predicates of x, P_n, being considered; if U has n members then $(\forall x)$ is defined as $\{P_1 \wedge P_2 \wedge ... P_n\}$ and $(\exists x)$ is defined as $\{P_1 \vee P_2 \vee ... P_n\}$. Each quantifier has a scope, marked by parentheses, as in $(x)(Sx \rightarrow Px)$ or $(\exists x)(Sx \wedge Px)$; these read "For every x, if x is an S then x is a P" and "At least one x is both an S and a P". Alternative readings are "All S are P" and "Some S are P", so that this is how this logic connects with traditional logic; we also have $(x)(Sx \rightarrow \sim Px)$ and $(\exists x)(Sx \wedge \sim Px)$, reading "No S are P" and "Some S are not P"; however the universal statements do not have existential presupposition (56) in modern logic.

There are four argument forms and one equivalence for use with quantified statements.

The equivalence is: $(x)(\Phi x) \equiv \sim(\exists x)\sim\Phi x$, where Φx represents any expression within the scope of the quantifier. From this and other truth-functional equivalences, other equivalences may be derived, such as $\sim(x)\Phi x \equiv (\exists x)\sim\Phi x$. The argument forms are:

Universal instantiation (U.I.): $(x)(\Phi x) \rightarrow \Phi a$.
Universal generalisation (U.G.): $\Phi a \rightarrow (x)(\Phi x)$.
Existential instantiation (E.I.): $(\exists x)(\Phi x) \rightarrow \Phi a$.
Existential generalisation (E.G.): $\Phi a \rightarrow (\exists x)(\Phi x)$.

Instantiation moves from a variable, x, to an ambiguous individual, a, and generalisation moves back again; these are needed because the truth-functional argument forms and equivalences apply to individuals, not classes of individuals. There are strict limitations on existential instantiation: it may instantiate only to an individual not previously introduced into the argument; and an existentially instantiated individual may not be generalised universally. An unambiguous individual is introduced only in a premise; for example, Ms for Socrates is mortal.

The syllogism Barbara, MaP, SaM, SaP, (see Appendix A) in modern logic is rendered:

1. (x)(Mx→Px)	Premise.
2. (x)(Sx→Mx)	Premise.
3. Ma→Pa	1, U.I.
4. Sa→Ma	2, U.I.
5. Sa→Pa	4, 3, Chain arg.
6. (x)(Sx→Px)	5, U.G.

and Ferioque, MeP, SiM, SoP, is rendered:

1. (x)(Mx→~Px)	Premise.
2. (∃x)(Sx∧Mx)	Premise.
3. Sa∧Ma	2, E.I.
4. Ma→~Pa	1, U.I.
5. Ma	3, Simplification.
6. ~Pa	4, 5, M.P.
7. Sa	3, Simplification.
8. Sa∧~Pa	6, 7, Conjunction.
9. (∃x)(Sx∧~Px)	8, E.G.

Note that in line 3 E.I. comes before U.I. in line 4 because of the limitation stated above.

The next illustration of a syllogism, Bramantip, PaM, MaS, SiP, requires an extra premise, line 3, to replace the existential presupposition of traditional logic:

1. (x)(Px→Mx)	Premise.
2. (x)(Mx→Sx)	Premise.
3. (∃x)Px	Premise.
4. Pa	3, E.I.
5. Pa→Ma	1, U.I.
6. Ma→Sa	2, U.I.

7. Pa→Sa	5, 6, Chain arg.
8. Sa	7, 4, M.P.
9. Sa∧Pa	8, 4, Conjunction.
10. (∃x)(Sx∧Px)	9, E.I.

The deficiencies of truth-functional logic (65) carry over into the monadic predicate calculus. For example, if something does not exist then whatever is said of it is true — as is easily shown. Let us assume that no mermaids exist: ∼(∃x)Mx. Then:

1. ∼(∃x)Mx	Premise
2. (x)∼Mx	1, Quantificational equivalence.
3. ∼Ma	2, Universal instantiation.
4. ∼Ma∨Qa	3, Disjunctive addition.
5. Ma→Qa	4, Implicative equivalence[32].
6. (x)(Mx→Qx)	5, Universal generalisation.

Line 6 says that every mermaid is a Q, where Q is anything you please. So if it is true that no mermaids exist then it is true that all mermaids are male, triangular, Danish, made of ice cream, and whatever else you fancy. This absurdity cannot be removed by simply denying Disjunctive Addition, because Boolean algebra (66) requires it.

Out of the monadic predicate calculus is developed the polyadic predicate calculus, in which a variable or individual may have more than one predicate. The word 'polyadic predicate' (104) was originally intended to make relations merely predicates, as in 'monadic predicate calculus' and 'polyadic predicate calculus'. For example, 'My cap is on my head' could be represented by Och, where O, 'on', has the polyadic predicates c and h, which stand for my cap and my head.

---

[32] (∼p∨q)≡(p→q), as is proved using a tree of truth-functional consequences, as in Fig. B1.

Further development occurs with modal logic, the logic of possibility and necessity. In this the possible is defined as anything that is true in at least one possible world, and the necessary as anything that is true in every possible world. Thus these definitions are based on the enlargement of the scope of the universal and existential quantifiers to every possible world; they have nothing to do with primitive possibility and primitive necessity.

Curiously, truth-functional modal logic leads to many, different, modal logics in an area where there should be only one.

Criticisms of modern logic are given in Chapter 3 (65).

Appendix C: Solipsism.

Solipsism (from the Latin, *sole ipse*, alone I am) is best understood with the concepts of **perceptible** and **imperceptible**. A perceptible, for me, now, is anything of which I am conscious, now; and an imperceptible, for me now, is anything not a perceptible for me now. 'Me' and 'now' refer to whomever is thinking about this, and when; in what follows the 'me' and the 'now' will be mostly implicit.

Solipsism then arises from the claim that no imperceptibles exist. That is, what I am conscious of now is all that exists. The following points are valid inferences from this premise:

1.   The past and the future are imperceptible so do not exist. I am conscious of some memories and some expectations (these are all such that exist) but they have to be all false, since past and future do not exist. (The past and memories of it have to be numerically distinct because memories can be false and the past cannot be false; similarly for the future and expectations.)

2.   So time does not exist. Only an eternal present, for me, here, now, exists. I am conscious of time passing, but since there is no time, this is an illusion.

3.   I am conscious of change, but since change requires time, this also is an illusion.

4.   I have beliefs (all and only those that I am conscious of now) about various imperceptibles, but since these imperceptibles do not exist all these beliefs are false.

5.   One of these beliefs is that all the empirical objects that I am conscious of now (which are all that exist) are real, in the sense of existing independently of being perceived. But since this belief is false, nothing empirical is real.

6.   I have explanations (all those of which I am conscious now) but since all explanations invoke imperceptible causes, all these explanations are false.

7.   One of these explanations is an explanation of why I find myself in this very peculiar situation of solipsism; this explanation is also false.

8.   Another explanation is that all other empirical people of which I am conscious now (there are no others) behave as they do because they have minds similar to mine; this explanation is also false.

9.   Some of these empirical people report to me of perceptibles for them, now, which are imperceptible to me; obviously, these reports are all false.

10.   I have a belief that the noumenal world exists, but this belief is false.

11. Thus everything of which I am conscious — everything that exists — is both inexplicable and either illusory or false.

        The only known cogent proof that some imperceptibles exist, so that solipsism is false, is the ontological argument.

Appendix D: Planck Units.

    Planck units, due to Max Planck (1858-1947), are units of measurement that are independent of arbitrary human standards. Length, for example, may be measured in units of length such as metres, miles, light-years, or parsecs, while various units of time are defined by the rotation of the Earth about its axis and about the Sun. Such units are human inventions and are thereby not part of the fundamental nature of the Universe. Planck units *are* fundamental, and free of human arbitrariness.

    In order to understand the Planck units we must first introduce dimensional analysis, a simple technique invented by James Clerk Maxwell[33] (1831-1879) and much used by engineers and scientists. The "dimensions" of dimensional analysis are not the three geometric dimensions of space or the four of space-time; rather, they refer to kinds of measurement; what might be called atoms of measurement, so that other kinds of measurements are structures made up of these atoms. Most commonly these dimensions are mass, length, and time, symbolised by M, L, and T. One example of this is speed, which might be measured in kilometres per hour. A kilometre is a length, L, an hour is a duration, T, and 'per' means divided by, so speed has the dimension L/T, more conveniently written as LT^{-1}. Another example is the flow of a liquid in, say, litres per minute; a litre is a volume, L^3, so the flow has dimensions $L^3 T^{-1}$. Energy has the same dimensions as work, which was defined by Newton as force times distance, force is mass times acceleration, and acceleration is change of velocity per duration, or LT^{-2}; so energy has dimensions $M(LT^{-2})L$ which simplifies to $ML^2 T^{-2}$. One advantage of dimensional analysis is that if you have an equation involving measurements then for the left hand side of the equation to equal the right hand

---

[33] Mahon, Basil, *The Man who Changed Everything: the Life of James Clerk Maxwell,* 2003, John Wiley and Sons.

side, each side must have the same dimensions; so analysing these dimensions is a convenient check on the correctness of the equation.

The main Planck units are derived from three fundamental constants of Nature, using dimensional analysis. The constants are the gravitational constant, G, the velocity of light, c, and Planck's constant, $\hbar$, the unit of action. G has dimensions $L^3 M^{-1} T^{-2}$, c has dimensions LT^{-1}, and $\hbar$ has dimensions $L^2 M T^{-1}$, which latter are the dimensions of energy times time, or of momentum times length. The Planck length, l_p, is defined as:

$$l_P = \sqrt{\frac{\hbar G}{c^3}}$$

and if you put in the dimensions for $\hbar$, G, and c in this you get:

$$\sqrt{\frac{L^2 M T^{-1} L^3 M^{-1} T^{-2}}{\left(LT^{-1}\right)^3}} = \sqrt{\frac{L^5 T^{-3}}{L^3 T^{-3}}} = L$$

which is as it should be because l_p is a length.

In similar fashion the Planck time, t_p, is defined as:

$$t_P = \sqrt{\frac{\hbar G}{c^5}}$$

of which the dimensions on the right simplify to T; and the Planck mass, m_p is defined as:

$$m_P = \sqrt{\frac{\hbar c}{G}}$$

which has the dimension M. If we define a maximum velocity as l_p/t_p then this is:

Appendix D: Planck Units.

$$\frac{\sqrt{\dfrac{\hbar G}{c^{3}}}}{\sqrt{\dfrac{\hbar G}{c^{5}}}} = c$$

The Planck units can be extended to bring in electromagnetism and thermodynamics by invoking the Coulomb constant and the Boltzmann constant.

In terms of the metre, kilogram, and second, $l_P = 1.6162 \times 10^{-35}$ metres, $m_P = 2.1765 \times 10^{-8}$ kilograms, and $t_P = 5.3911 \times 10^{-44}$ seconds.

Appendix E: Duality.

In orthodox set theory intersection and union are **duals** of each other; and so are the universe of discourse, U, and the null set, ϕ. This means that any true expression containing these can yield a dual expression with each element replaced by its dual, and this dual expression is also true. For example, the expression $(S \cap P) \cup (S \cap P') = S$ can be proved to be true and so, by duality, $(S \cup P) \cap (S \cup P') = S$ is true; and since $S \cap U = S$ is true, so is $S \cup \phi = S$.

The principle of duality in mathematics is the principle that if any axioms imply their own duals then any theorems implied by these axioms also imply their own duals.

In Chapter 4 (81) we saw duals between the relational connectives and the extensional connectives. Another example of duality occurs in projective geometry, in which points and lines in the projective plane are duals of each other, so any true theorem about points on lines implies a true theorem about lines through points, and *vice versa*. The simplest example of this is the Fano plane, shown in Fig. E1, where a point is shown as a black dot. This plane has seven lines (including the circle) and if you observe that each line has three points on it then you can immediately infer that each point is on three lines. Or the other way round.

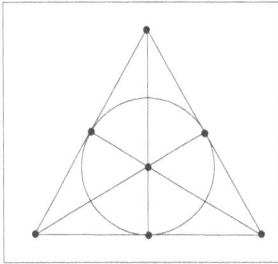

Fig. E1.

Bibliography.

Davies, P. C. W., and J. R. Brown, *The Ghost in the Atom: A Discussion of the Mysteries of Quantum Physics*, Cambridge, C.U.P., 1986/1993.

Leibniz, *Monadology*.

Mahon, Basil, *The Man who Changed Everything: the Life of James Clerk Maxwell*, 2003, John Wiley and Sons.

Plato, *Parmenides*, *Phaedo*, *Republic*, *Symposium, Timaeus*.

Robinson, Helier J., *Renascent Rationalism*, Toronto, MacMillan of Canada, 1975; 5th. ed., Sharebooks Publishing, www.sharebooks.com, 2008.

Russell, Bertrand, *A History of Western Philosophy*, Allen and Unwin, London, 1946
   Human Knowledge, Its Scope and Limits, London, George Allen and Unwin, 1948; Routledge ed., Oxford, 2009.
   Mysticism and Logic, London, George Allen and Unwin, 1918.

Schrodinger, Erwin, *What is Life?*, Cambridge, C.U.P., 1944.

Spinoza, *Ethics*.

Stewart, Ian, *Why Beauty is Truth: A History of Symmetry*, New York, Basic Books, 2009.

Voltaire, *Candide*.

Whitehead and Russell, *Principia Mathematica*, Cambridge, C.U.P., 1910.

Wigner, Eugene, *The Unreasonable Effectiveness of Mathematics in the Natural Sciences*, in *Communications in Pure and Applied Mathematics*, vol. 13, No. 1 (February 1960). New York: John Wiley & Sons, Inc.

Glossary of Symbols.

Fonts:

Times New Roman

U.C: relations, as in R, S.

U.C. italic: sets, as in *P*, *Q*.

L.C: propositions, statements, as in p, q, r.

Small caps: terms of relations, as in R, S.

Small caps, italic: sets of terms of relations, as in *R*, *S*.

Small caps, capped: properties of relations, as in R̂, Ŝ.

Small caps, capped, italic: sets of properties of relations, as in *r̂*, *ŝ*.

Subscripts: denotation of particular individuals and scopes. The scopes, which apply to quantifiers and to uniqueness, are R, W, U, and P̂; R standing for all relations in all possible worlds, W standing for all of a possible world, U standing for all possible worlds and P̂ standing for all possible properties of a relation.

Reserved letters:

M̂ : the minim: the three properties of adicity, simplicity, and possibility.

W: a whole.

W: a possible world.

C: a consciousness.

T: the top relation of a whole or of a possible world.

Arial: Reserved letters:

G: the best of all possible worlds.

T: the top relation of G.

U: unique. Subscripts denote the range, as above under "Subscripts".

Truth-functional logic: (231).

$\sim$: not.

$\wedge$: and.

$\vee$: or.

$\rightarrow$: material implication.

$\equiv$: equivalence.

Quantificational logic: (234)

(x): every x, as (x)(Px$\rightarrow$Qx): every P is a Q.

($\exists$x): at least one x, as ($\exists$x)(Px$\wedge$Qx): at least one P is a Q.

Set theory: (69).

$\{\mathcal{E}p\}$: the set of every p.

$\langle a,b \rangle$: the ordered set having a as its first member.

ϵ: set membership, as xϵS: x is a member of the set *S*.

$\notin$: non-membership, as in x$\notin$S: x is not a member of the set *S*.

$\ni$: reads 'such that'.

U: universe of discourse.

ϕ: the null, or empty, set.

=: set identity, as: the sets *P* and *Q* are identical, $(P=Q)\Leftrightarrow(x)[(x\epsilon P)\Leftrightarrow(x\epsilon Q)]$.

$\subset$: proper subset, as $P\subset Q$: *P* is a proper subset of *Q*: $(P\subset Q)\Leftrightarrow(x)\{[(x\epsilon P)\Rightarrow(x\epsilon Q)]\wedge\sim(P=Q)\}$.

$\supset$: proper superset, as $P\supset Q$: $(P\supset Q)\Leftrightarrow(x)\{[(x\epsilon Q)\Rightarrow(x\epsilon P)]\wedge\sim(Q=P)\}$.

$\subseteq$: subset, as $P\subseteq Q$: *P* is a subset of *Q*, $(P\subseteq Q)\Leftrightarrow(x)[(x\epsilon P)\Rightarrow(x\epsilon Q)]$.

$\supseteq$: superset, as $P\supseteq Q$: *P* is a superset of *Q*, $(P\supseteq Q)\Leftrightarrow(x)[(x\epsilon Q)\Rightarrow(x\epsilon P)]$.

$\cup$: union, as $P\cup Q$: the union of *P* and *Q*, $(P\cup Q)\Leftrightarrow(x)[(x\epsilon P)\vee(x\epsilon Q)]$.

$\cap$: intersection, as $P\cap Q$: the intersection of *P* and *Q*. $(P\cap Q)\Leftrightarrow(x)[(x\epsilon P)\wedge(x\epsilon Q)]$.

P': negation, non-*P*: $(x)\{(x\epsilon P')\Leftrightarrow[x\epsilon(U\text{-}P)]\}$.

Relational logic:

P̂, Q̂, R̂, etc. (45): properties of a relation, as in P̂R.

P̂, Q̂, R̂, etc. (45): property sets of a relation, as in P̂R.

R^s or R^s (45): upper extrinsic properties of R.

^{n}R (91): the adicity, *n*, of a relation R.

Ɛ (46): the relation-every.

A (46): the relation-any.

(*Ɛ*R), (*Ɛ*R̂), (*Ɛ*X), and (*Ɛ*x) (46): relational universal
quantifiers; R̂, R̂, X, x, etc. denote the universe
of discourse of the quantifier.

≋ (75): similarity, as in P̂≋Q̂ or P̂≋Q̂.

ι (53): true, as in ιp; normally omitted.

≈ (75): dissimilarity, as in P̂≈Q̂ or P̂≈Q̂; see also
(≋ŝ)′.

~ (53): false, as in ~p.

⅄ (73): commonality, as in P̂⅄Q̂.

Y (73): coupling, as in P̂YQ̂ or P̂YQ̂.

≻ (72): superintension, as in P̂≻Q̂.

≺ (72): subintension, as in P̂≺Q̂.

⋀ (74): polyadic disjunction, as ⋀ŝ.

(≋ŝ)′ (75): the relational complement of ≋ŝ, also
written ≈ŝ.

⇒ (47): necessity and implication, as in P̂⇒Q̂.

⇐: (47) necessitated by, as in P̂⇐Q̂.

⇔ (47): binecessity and equivalence, as in P̂⇔Q̂.

⇑ (47): upward, or bottom-up, necessity.

⇓ (47): downward, or top-down, necessity.

⇓⇑ (135): loop necessity.

#R̂ and #R, or #R̂R. (47): the emergent levels of R̂ and
R.

⇅ (47): inherence.

R⇒E⇒N (41): the relation between relational,
extensional, and nominal meanings.

◊ (45): intrinsic possibility.

◊̂ (45): extrinsic possibility.

⊡ (45): intrinsic necessity.

◌̇ (45): extrinsic necessity.

#R (21): the lowest level at which R can emerge.

Â (134): in Ch. 9, actuality, a property of a relation.

Φ (45): any logic formula.

U (146): unique. U_R stands for unique among all relations in all possible worlds, U_W stands for unique among all of a possible world, U_U stands for all unique among possible worlds and $U_{\hat{p}}$ stands for unique among all possible properties of a relation.

$U_{\hat{p}}$ (76): the universe of discourse of all intrinsic properties.

U_R (76): the universe of discourse of all kinds and all instances of relations.

$U_{\hat{p}\,\succ\,\hat{M}}$ (76): the universe of discourse of all kinds of relations.

Glossary of Defined Terms.

Words printed in italic font are also entries in this Glossary. Parenthetical numbers are cross-references to the page number of the original introduction of the symbol.

absolute value (37): of a *relation* R is its *hekergy*: if the number of *emergence-configurations* of the *term set* of of R, out of which R *emerges* is *e*, and the total number of possible configurations is *t* then the hekergy of R is ln.(t/e). The numbers *e* and *t* are both finite.

abstract (13, 24): anything that has no *concrete* qualities; anything that cannot be *imagined* in isolation. In this book, all and only *relations* and their *properties*.

abstract idea (55): a *relation* or *property* of a relation in the mind; an *ideal relation* or ideal property.

abstraction (188): a special kind of *mapping*, as are *discriminatory mappings*. A mapping is a *relation*.

abstract thought (55): manipulation of *abstract ideas* and discovery of their abstract properties and of *relations* between them. See also *pure thought, conceptual thought* and *nominal thought*.

actual existence (43): the existence possessed by the *empirical world* and by the *noumenal world*; *intrinsic necessity*.

actuality (134): *intrinsic necessary existence* of a *relation*, symbolised by ⊡. A relation is actual if its intrinsic necessary existence is necessary, as happens with ⊡T, or else its intrinsic necessary existence is *extrinsically necessitated*, ⨎⊡, by *inherence, emergence,* or *demergence*. If T exists in a possible world then that world has loop actuality and so is actual.

Actuality is also *empirical existence*, which is best defined, stipulatively and personally, by Descartes' *cogito ergo sum* (I am conscious therefore I exist); we all know actuality empirically.

addition (91, 92): of two *relational numbers*, n and m, symbolised by $n+m$, is their *coupling*: $(n+m)=(\hat{N}\curlyvee\hat{M})$. This is called the **sum** of n and m. Addition may also be a *join*, symbolised by +, between the *terms* of an n-adic *relation*, R, each term being a unity and so a *representative instance* of the number one, such that $1_1+1_2+...1_n = n$. The addition of two *relational numbers* m and n is also the relational binary operation, or function, having the set of the two of them as its argument and their sum as its value. Addition is a *compoundable relation*. If we define the *subordinate terms* of a relation R as the terms of the terms of R, the terms of their terms, and so on down to the lowest level, such that the first level subordinate terms of R are the terms of the terms of R, then we can also define the first level subordinate adicity of a relation R as the sum of the adicities of the terms of R. So if R has the m terms $^aR_1, {}^bR_2, ... {}^nR_m$, where the superscripts are their adicities, then the first level subordinate adicity of R is $a+b+...n$.

adicity (16): the number of *terms* that a *relation* possesses; an *intrinsic property* of every relation; one of the three *properties* in the *minim*, the others being *simplicity* and *possibility*.

agent (188): anything that has some awareness of its environment and some control over it, as with the *noumenal ego*.

analytic truth (54, 60, 63): may be relational, extensional, or nominal. The first is *superintension* and hence based on *necessity*; the second is *subset* and hence based on *universality*; the third is nominal *consistency* and hence based on absence of contradiction.

angle separator (100): speculatively, a *separator* that is a right angle and is symmetric.

astrology (198): the mistaken attribution of *oge* influences to the empirical planets.

atomic duration (38, 101): a *temporal separator*; a *compoundable relation* such that a chain of atomic durations, demarcated by *dissimilarities*, has a compounded duration.

atomic length (38, 100): a *linear separator*; a *compoundable relation* such that a chain of atomic lengths, demarcated by *dissimilarities*, has a compounded length.

atomic noumenal idea (181, 288): a *noumenal* neural On or Off in the *noumenal brain*.

atomic square (100): the simplest level-*2 whole*; also the *top relation* of the *simplest possible world*. It is composed of four *atomic lengths* separating four *angle separators*, with a *dissimilarity* between each length and angle.

atomic vector (101): speculatively, a *separator* that has length, like a *linear separator*, but a length slightly less than that of a linear separator; so that a *structure* of them forming an atomic volume has unit mass, through stressing space-time according to an inverse square law. This requires that separators are elastic and so can transmit gravitational waves. An atomic vector has a sense and so is asymmetric.

attention (188): the *ego* can distort itself in order to focus on what is of interest to its *selfish goals*. This focussing is its attention.

axiom generosity (40): the wealth of definitions and theorems that a good axiom set produces; it results from *cascading emergence* of *relations*; it is characteristic of mathematics and of the *Grand Structure*.

axiom level (24): the lowest *level* of all *relations* in a *possible world*; also called *level-1*.

axiom relations (24): *separators* in the *axiom level, level-1*.

belief (188): the difference between the *ego* considering a *proposition* and believing it is that in considering it the ego is simply conscious of it, while in believing it, the ego incorporates the belief into itself, as it does a *memory*.

best of all possible worlds (38): the *possible world* that has the highest *quality* of any possible world. The quality is the total amount of emergent *hekergy* in a *possible world*, divided by the number of *separators* (*axiom relations*) in its *axiom level*. (Axiom separators have no hekergy.) It is proportional to the *level* of the *top relation* of the world See also *B.P.P.*

bonding (189): two or more *noumenal ideas* can become permanently attached to each other, as in the bonding of an *abstract idea* to a word to form an *abstract concept*, or in the bonding of a *motor idea* with the desire of a particular muscular movement.

boundary-intension (50): an *intension* that defines a *set* by means of a *boundary*; it is a relation B such as in, on, or bounded by, which has the boundary as one of its terms and any possible relation within the boundary as its other term. The set defined by a boundary-intension is called a *boundary set*

boundary set (50): A *set* defined by a *boundary intension*. Examples are 'Everything in this box', 'All the people in this room', 'All the food on the table' and 'All the numbers between 1 and 100'.

binecessity (60): symmetric *necessity*. Symbol: ⇔.

Boolean algebra (66): a two-valued algebra, also variously called truth-functional algebra and truth-functional calculus, which forms the basis of modern symbolic logic, or *truth-functional logic*.

bottom up necessity (21): *upward necessity*; the *necessity* of *emergence*.

boundary (33): a *dissimilarity relation* or a contiguous series of them.

B.P. (143): abbreviation for the *best of all possible worlds*.

B.P.P. (143): abbreviation for the principle that the *best of all possible worlds* exists necessarily because it is the best.

Cartesian product (5, 84) of two sets, P and Q, symbolised $P \times Q$, is: $(P \times Q) = \{\langle x,y \rangle : (x \in P) \wedge (y \in Q)\}$, where the angle brackets $\langle , \rangle$, signify an *ordered pair*. The definition may be expanded, as in $(P \times Q \times ...S) = \{\langle w, x, ...z \rangle \ni [(w \in P) \wedge (x \in Q) \wedge ...(z \in S)]\}$.

cascading emergence (25): the *emergence* of a series of *relations*, at successively higher *levels*. See also the *Grand Structure*.

categorical propositions (224): the basic propositions in traditional logic. There are four kinds: "All S are P" (universal affirmative proposition), "No S are P" (universal negative proposition), "Some S are P" (particular affirmative proposition), and "Some S are not P" (particular negative proposition. 'S' stands for subject and 'P' for predicate.

categorical syllogism (226): an *argument* in *traditional logic* using two *categorical propositions* as premises and a third as conclusion. See also *major premise* and *minor premise*.

causation (7): a *noumenal necessity* which occurs in *parallel* with a *relation* of duration. It is the basis of explanation: to describe causes is to explain their effects; and causation is never empirical, it is always noumenal — hence our need for explanations leads us to postulate the *existence* of the *noumenal world*, which is all that both exists and is imperceptible.

change (33): a relation of *dissimilarity* in *parallel* with a *duration*.

closed loops (31): *separators* in the axiom level are characterised by forming closed loops; because if they did not close then they would have to end, and the end separator would be missing a term and so could not exist. See also *loop possibility* and *loop necessity*.

coherence truth (105): richness of *necessities* in an *explanation*; or, equivalently, absence of arbitrariness.

commonality (73): the *relation* between two *property sets*, $\hat{s}$ and $\hat{r}$, symbolised by $\hat{s} \curlywedge \hat{r}$, is such that each member of $\hat{s} \curlywedge \hat{r}$ is an improper subintension of both a *member* of $\hat{s}$ and a member of $\hat{r}$. If the commonality of $\hat{s}$ and $\hat{r}$, other than the three universal relational intrinsic properties of possibility, simplicity, and adicity (the *minim*) does not exist, $\hat{s}$ and $\hat{r}$ are said to be *disparate*.

complete disjunction (76): a relational *commonality*, or disjunction, whose *extension* is an *intensional set*. If the extension is a *contingent set* the disjunction is incomplete and not regarded as a disjunction, relationally. See also the *disjunction theorem* and the *relational disjunctive addition theorem*.

compoundable relation (38): any *kind* of *relation* that has one or more *properties* which are also possessed by the relation which unites a set of *instances* of that kind.

compounded relation (38): any *kind* of *relation* that has one or more *properties* which are also possessed by its terms. See also *definite compounded relations*.

concept (55): an *abstract idea* bonded to a word or symbol. Also sometimes refers to a *concrete concept*, which is a concrete image bonded to a word.

conceptual thought (56): thought with *concepts*, which are *abstract ideas* bonded to words or symbols. Compare with *pure thought*, which is thought with abstract ideas alone; *extensional thought*, which is thought with *extensions*, and *nominal thought*, which is thought with words or symbols alone.

concrete (13, 24): any sensory quality, such as colours, sounds, smells, tastes, and degrees of hot and cold, rough and smooth, hard and soft, penetrable and impenetrable, light and heavy, etc. Anything perceived or imagined other than relations and their properties.

concrete concepts (56): words *bonded* to *concrete images*.

concrete image (56): any *image* in the *imagination*.

configuration (39): the set of all of the *separators* between the *terms* of a *relation*. Configurations are either *emergence-configurations* or *submergence-configurations*.

conjunction (42, 73): may be *coupling*, which is its *relational meaning*, or *intersection*, which is its *extensional meaning*. It also has nominal meaning in truth-functional logic, where it is defined as the connective that is true if both conjuncts are true, and is otherwise false.

conjunction theorem, the (79): $\{\mathcal{E}(\hat{A} \curlyvee \hat{B})\} = \{\mathcal{E}\hat{A}\} \cap \{\mathcal{E}\hat{B}\}$.

connectives (69): In logic, connectives are *relations* that combine *concepts* into other concepts or into *propositions*, or else propositions into larger propositions. In *truth-functional logic* they combine statements into larger statements.

consciousness (183): the reaction within the *noumenal ego* from the force of interaction between the noumenal ego and any noumenal idea; this reaction is *empirical*. A *mid-sensation* produces the consciousness of an *empirical sensation*, a *mid-object* produces the consciousness of an *empirical object*, and a *mid-world* produces the consciousness of an *empirical world*.

consistency (17): *possibility*, a primitive and *intrinsic property* in the *minim*.

contingency (42): with *necessity* and *impossibility*, contingency is one of the three traditional *modalities*; a plural *possibility*, as opposed to *singular* and *zero possibility* which are *necessity* and *impossibility*.

contingent function (111): a *contingent set* of assignments of a unique value to each member in an *extensional set* called the domain of the function. See also *relational function* and *nominal function*.

contingent possible world (142): a *possible world* in which at least one relation is *contingent*.

contingent set (29, 40): a *set* which has no *intension*. It can only be defined by *enumeration* of its *members*, so if it is too large to enumerate it has *nominal meaning* only; its *membership* is the *demergence* of a *term* of a *set relation*.)

contingent set theorem (79): an *extensional S* is a *contingent set* if and only if $S \subset \{\mathcal{E}(\wedge S)\}$.

contingent set theory (84): set theory that excludes *intensions*. Also called *exclusively extensional set theory*.

conversion (225): in *traditional logic*, an *immediate inference* from a *categorical proposition* that interchanges subject and predicate, provided that an *undistributed term* does not become *distributed*. See also *obversion*.

correspondence truth (105): *similarity* to *reality*.

coterms (19): if a *relation* R is a *term* of relation S then the *set* consisting of S and its terms,, is an upper extrinsic property, or *upper*, of R. These terms of S other than R are called the coterms of R in S.

coupling (73): the coupling of two *intrinsic property sets*, $\hat{s}$ and $\hat{r}$, symbolised by $\hat{s} \curlyvee \hat{r}$, is such that each *member* of $\hat{s} \curlyvee \hat{r}$ is an improper superintension both of $\hat{s}$ and of $\hat{r}$.

creativity (189): increase of *hekergy*, in the *empirical mind* or *empirical world*, in each of which it is a *consciousness* of hekergy increase in the *noumenal mind*.

decoupling (74): of two *intrinsic property sets*, $\hat{s}$ and $\hat{r}$, which are not *disparate*, symbolised $\hat{s}$-$\hat{r}$, if it exists, is the *set* consisting of those *members* of $\hat{s}$ which are not *similar* to any member of $\hat{r}$.

definite compounded relations (38): chains of *similar compoundable relations* that have *dissimilarity relations* at each end.

degree of dissimilarity (39, 182): given any two *relations*, if they have a number, s, of *similar intrinsic properties* and a number, d, of *dissimilar* properties, and one has m more properties than the other, then the degree of dissimilarity between them is $(d+m)/(s+d+m)$, and the *degree of similarity* between them is $s/(s+d+m)$.

degree of similarity (39, 182): given any two *relations*, if they have a number, s, of *similar intrinsic properties* and a number, d, of *dissimilar* properties, and one has m more properties than the other, then the degree of similarity between them is $s/(s+d+m)$, and the *degree of dissimilarity* between them is $(d+m)/(s+d+m)$.

demergence (20): the *necessary* coming into *existence* of the *terms* of a *relation*, R, given the existence of R, since a relation cannot exist without its terms.

demergence-intension (50): is a *demergence*, D, and its *relation*, R: $\Downarrow$R. It defines the *term set* of R — all the *terms* that demerge from R: $\{\mathcal{E}\Downarrow R\}$.

depression (197): a *feeling* in the *ego* caused by oge-oppression.

design (21): a process of human *creation* of *wholes* with specific *emergent intrinsic properties* in their *top relations*.

This is *downward necessity* from the desired emergents to the parts.

disjoin (42): either the *submergence* or the non-existence, of a *join*; a join being the relation with the simplest possible *intrinsic property set*, consisting of only the *minim* — that is, the properties of *possibility*, *simplicity*, and *adicity*.

discrimination (189): the ego is capable of special mappings that enable it to discriminate. For example, a *boundary* is a series of contiguous *dissimilarities*; if just this is closed and mapped, the result is a shape. A scale mapping, up or down, to *similarity*, gives relative size. And a mapping of a small patch gives *concrete* qualities such as colour.

disjoint (74): If the *intersection* of sets S and T does not *exist*, S and T are said to be disjoint.

disjunction (42, 73,): may be *coupling*, which is its *relational meaning,* or *union*, which is its *extensional meaning*, or nominal meaning in truth-functional logic. Except for its relational meaning, all these meanings are defined as a *connective* which is false if both disjuncts are false, and is otherwise true. Disjunction may also be a *disjoin*, which is the absence of the *relation* called a *join*, which is the kind of relation having the *simplest intrinsic property* set, the *minim*; this applies to individuals.

disjunction theorem, the (80): $\{\mathcal{E}(\hat{A} \wedge \hat{B})\} \supseteq \{\mathcal{E}\hat{A}\} \cup \{\mathcal{E}\hat{B}\}$.

disjunctive addition (61): a *truth-functional* valid argument form that is valid in *extensional logic* but is trivial in *relational logic*: $p \rightarrow (p \vee q)$ and $P \subset (P \cup Q)$. The value of this argument form is questionable because it allows the introduction of unlimited irrelevance, q or Q, into an argument. Its triviality in relational logic, is shown by

Theorem 4.10, the *disjunctive addition theorem*:
$\{\mathcal{E}\hat{P}\}\subset[\{\mathcal{E}\hat{P}\}\cup\{\mathcal{E}\hat{Q}\}]$ which requires that $\{\hat{s}\Rightarrow\{(\hat{s}\curlywedge\hat{P})$.

disjunctive addition theorem (80, 208): $\{\mathcal{E}\hat{P}\}\subset[\{\mathcal{E}\hat{P}\}\cup\{\mathcal{E}\hat{Q}\}]$ if and only if the disjunction is complete, which is true if and only if $\{\mathcal{E}\hat{A}\,R\}\cup\{\mathcal{E}\hat{B}\,R\}$ is an intensional set.

disparate (73): if the *commonality* of $\hat{s}$ and $\hat{t}$, $\hat{s}\curlywedge\hat{t}$, other than the three universal *intrinsic properties* of *possibility*, *simplicity,* and *adicity* (the *minim*) does not exist, $\hat{s}$ and $\hat{t}$ are said to be disparate.

dissimilarity (33): a *relation* which may be empirical, ideal, or noumenal; it would multiply endlessly were it not restricted (i) by not having any *uppers* and (ii) by having only locally separated *terms*. It is necessary in the definitions of a *boundary* and a *change*. It also emerges in higher level structures such as minds or computers, as a result of comparisons. Symbol: $\approx$. See also *similarity*.

dissimilarity falsity (53): if a *relation*, *whole*, or *structure* is a copy, representation, *image*, or reproduction of another, and they are *similar*, then their similarity is called the *relational truth*, or *similarity truth*, of the copy, relative to the other, or original; if they are not similar then the copy is *relationally false*, or dissimilarity false, relative to the original. Also called *relational falsity*. Generally the original is either *empirical* or *noumenal* and the image is *ideal*.

dissimilarity-intension (49): a *dissimilarity* and one of its terms, R, say: $\approx$R. This defines a *dissimilarity-set*: $\{\mathcal{E}\approx R\}$ or, more fully, $\{\mathcal{E}X: X\approx R\}$. This set is the set of every relation that is dissimilar to R. $\{\mathcal{E}\approx R\}$ is then the complement of $\{\mathcal{E}\wr R\}$, or $\{\mathcal{E}\wr R\}'$, the set of every non-R.

dissimilarity set (75): if {Ɛℛŝ} is a *similarity set* then {Ɛ≈ŝ)} is its corresponding *dissimilarity set*. {Ɛ≈ŝ)} is the *extensional complement* of {Ɛℛŝ} and ≈ŝ is the *relational complement* of ŝ.

distribution (224): in *traditional logic* a *term* is distributed in a *categorical proposition* if, in that proposition it refers to the entire class named; and otherwise it is *undistributed*.

divine right of kings (198): monarchs used to believe that they had a divine right to rule because their coronation was a process of approval thereof, by God. The *oge* is one of the meanings of the word god, thus a coronation is a divine *rite of passage*.

downward necessity (21): *top down necessity*; *demergence*; the *necessity* which makes the *existence* of a *relation* require the existence of its *terms*.

dual (81): See Appendix E (244).

ego (183): All the *noumenal memories* that form the ego have a major component in common, which is the *mid-memory* of the *mid-body* of the person concerned, the mid-body being an image of her or his *noumenal body*, of which latter the *noumenal brain* is a part. This common component makes these mid-memories largely *similar*, so that they mutually attract, by *L.A.L.*, and form a *structure* which is the noumenal ego.

ego-dominant (196): an ego-dominant person is more *selfish* than *moral* because the ego is dominant over the *oge*. See also *oge-dominant*.

elation (197): a *feeling* felt by the *ego* as it oppresses the *oge*.

emergence (14): the coming into *existence* of a property set, P̂, that is necessitated by an *emergence-configuration* of the terms of P̂R. Often abbreviated to: "The emergence configuration emerges R" rather than "The emergence configuration emerges the intrinsic properties of R, which inhere R".

emergence-configurations (20): are those *configurations* of the *terms* of a *relation* which *necessitate* the *emergence*, or else continued *existence*, of the *properties* of that relation. See also *submergence-configurations*.

emergence-intension (50): C⇑; an *emergence*, ⇑, and an *emergence-configuration*, C, of the *term set* of a *relation*, R, which has the *property set P* of R; *P* is what emerges out of C and *inheres* (necessitates) the existence of R.

emergent level (21): the lowest *level* of a *structure* or *possible world* at which an *intrinsic property* or *relation emerges*. The emergent level of a particular *kind* of *relation* is the same in all *possible worlds* in which that kind of relation *exists*. Symbol: #R̂R or #R or #R̂.

emergent separators (100): *separators* that *emerge* above the *axiom level*; they separate other separators, such as *atomic areas,* but they are not *atomic separators*, like *axiom separators,* because they do not have to form *separator loops*.

empirical actuality (135, 137): everything empirical that exists. See also *empirical reality*.

empirical causation (126): constant conjunction of contiguous and successive *empirical* events (David Hume); high empirical correlation.

empirical existence (135): is best defined, stipulatively and personally, by Descartes' *cogito ergo sum* (I am conscious therefore I exist); it is also *actuality*, which we all know empirically.

empirical falsity (105): *empirical* illusion.

empirical memories (184): explained as *consciousness* of *mid-memories*.

empirical object (184): explained as *consciousness* of a *mid-object*.

empirical perception (191): an image of *noumenal perception*, experienced as a given.

empirical reality (53): all that exists in the *empirical world* that is not illusory. See also *empirical actuality*.

empirical relations (26): *relations* in the *empirical world*, relations known through the senses.

empirical sensation (184): explained as *consciousness* of a *mid-sensation*.

empirical truth (105): non-illusion, as opposed to *empirical falsity*, which is *illusion*.

empirical world (7, 11, 27, 184): the world that we know through the senses; all that exists and is perceptible; explained as *consciousness* of a *mid-world*.

endless multiplication (14): a feature of some *relations*, such as *self-similarity*, which, if it is a real entity, applies to itself; thus an instance of self-similarity is self-similar, and this second instance of self-similarity is self-similar, and so on, without end. We say that such relations have neither *empirical* nor *noumenal existence* because of *Occam's Razor*: they multiply entities beyond necessity because they explain nothing.

enumeration (50): an enumeration of a *plurality* is a list of the names, symbols, or descriptions of every *element* of that plurality; as such it is a serial dyadic *join* of names, symbols, or descriptions.

equal (91): Two *relational natural numbers* m and n are equal, symbolised $m = n$, if their adicities are similar, or if they have *representative instances* M and N which are *equiadic*. Two sets have equal *extensional numbers* of members if they are in *one-to-one correspondence*.

equiadic (91): two *relations* are equiadic if their *adicities* are *similar*.

equivalence (57, 58): a *relation* having the seven *properties* of *necessity*, symmetry, reflexiveness, and transitivity, as well as the properties in the *minim*: *adicity*, *possibility*, and *simplicity*.

equivalence theorem, the (79): $(\hat{P} \wr \hat{Q}) \Leftrightarrow (\{\mathcal{E}\hat{P}\} = \{\mathcal{E}\hat{Q}\})$.

esse est percipi (190): the claim that to be is to be perceived; that is, everything empirical exists only for as long as it is perceived.

evil (205): *hekergy* loss, as *perceived empirically*, subjectively, and selectively.

exclusively extensional function (111): an *extensional function* that is a *contingent function*.

exclusively extensional set (110): a *set* that has an *extension* but no *intension*; also called a *contingent set*.

exclusively extensional set theory (87): set theory that excludes *intensions*. Also called *contingent set theory*.

exclusively nominal function (112): a *nominal function* that has neither relational nor extensional meaning, such as a function whose domain, range, or both, are either null or infinite.

exclusively nominal relations (16): *relations* that can be named and/or described, but the names and descriptions have no reference. Also called *purely nominal relations*.

exherence (47): the inverse of *inherence*: *relations* exhere their *intrinsic properties*, and these inhere their relations.

exherence-intension (49): an *exherence*, E, and the *term*, R, which exheres the *properties*, $\hat{r}$, of a *relation*, R: $\uparrow\downarrow$R. It defines the *property set* of R, $\{\mathcal{E}\uparrow\downarrow R\}$, hence $\{\mathcal{E}\uparrow\downarrow R\}=\hat{r}$.

existence of a relation (17): is *mathematical* or *logical existence*, which is *intrinsic possibility*. It may also be *actual existence*, which is intrinsic singular possibility, or *intrinsic necessity*, which is the existence of the *empirical world* and of the *noumenal world*.

existential presupposition (56, 229): the claim that whatever is being discussed exists, or else the supposition that if a *universal proposition* SaP or SeP is true then the *sets S* and *P* have *members*. Applies in traditional and *relational logic* but not in modern *truth-functional* quantificational logic or set theory.

explanation (7, 10): description of *causations*: to describe causes is to explain their effects.

explanatory value (28): the criterion for qualification as *noumenal*; the better a theory is at explaining, the more probable is the *noumenal existence* of its content. The criteria of good explanation are discussed in Chapter 7.

extension (3): of a *set* is the *plurality* that, when unified by the *relation-every*, is the set; the totality of the *members* of a set.

extension theorem, the (78): for any set S, $S \subseteq \{\mathcal{E}(\wedge S)\}$.

extensional analytic truth (60): subset.

extensional causations (111): empirical high correlations. Also called *empirical causations*.

extensional complement (76): if U is the universe of discourse, having every *relation* as its *extension*, and S is an *extensional set* of relations then the extensional complement of S, symbolised S', is $S' = U - S$. S' is such that $S \cup S' = U$ and each member of S is not identical with each member of S', and vice versa. S' also reads as non-S. See also *relational complement*.

extensional conjunction (74): *intersection*.

extensional connectives (71): *connectives* between *extensions* which are defined by means of *identity*; see also *relational connectives*, which are defined by means of *similarity*.

extensional contradiction (61, 105): the null set. Because one form of *extensional truth*, called *extensional precursor truth*, is *set-membership*, the null set is always false. See also *extensional tautology*.

extensional definition (10): is definition of words that have
extensional meaning.

extensional disjunction (73): *union*.

extensional equivalence (72): *set identity*, more commonly
mis-called *set equality*; it is *extensionally valid* if and only if
either (i) membership in A is universally membership in B, and
vice versa: always if $x \in A$ then $x \in B$, and if $x \in B$ then $x \in A$, or
(ii) non-membership in B is universally non-membership in A,
and *vice versa*: always if $x \notin B$ then $x \notin A$, and if $x \notin A$ then $x \notin B$.
See also the *equivalence theorem*.

extensional function (111): a set of ordered pairs determined
by either a *relational function* or a *contingent function*, as
opposed to a *relational function* which is any *necessity*, or a
contingent function which is a *contingent set* of assignments of
unique values to every *member* in an *extensional set* called the
domain of the function, or a *nominal function* which is an
extensional function whose domain, range, or both, are
nominal sets.

extensional implication from A to B (60): is $A \subset B$, and is
extensionally valid if and only if either (i) *membership* in A is
universally membership in B: always if $x \in A$ then $x \in B$; or (ii)
non-membership in B is universally non-membership in A:
always if $x \notin B$ then $x \notin A$. Thus extensional validity is based on
extensional necessity (universality), not on *relational necessity*
(*singular possibility*).

extensional logic (62): the logic of *extensional sets*: a logic of
sets which may or may not have *intensions*; that is, it includes
both *necessary sets* and *contingent sets*.

extensional meaning (10, 40): a symbol, name, or description
has extensional meaning if its meaning is an *extension*; and it

has exclusively extensional meaning if it has extensional meaning but no *relational meaning*. See also *nominal meaning*.

extensional meaning of a relational number *n* (98): the *term set* of an *n*-adic *relation*, which is an *n*-membered *intensional set*.

extensional necessity (62, 106): universality, as opposed to *relational necessity* which is singular possibility, and *nominal necessity* which is truth-functional tautology. See also the *problem of induction*.

extensional number *n* (5, 98): the set of all *n*-membered *extensional sets*; that is, the set of all sets, *necessary* or *contingent*, that are in *one-to-one correspondence* with an *n*-membered extensional set.

extensional precursor truth and falsity (61): set membership and non-membership.

extensional relation (5): any *Cartesian product* or *subset* thereof, which is to say an *ordered set*. Because there are no ordered one-membered sets there are no *monadic* extensional relations.

extensional relation-any (111): a *contingent function* which yields a random member of a *contingent set*; also, the existential quantifier of quantificational logic.

extensional relation-every (111): a *contingent function* which is the *extensional meaning* of the *relation-every, &*; also the universal quantifier of quantificational logic.

extensional set (29): a *set* which is either a *necessary set* or a *contingent set*.

extensional set theory (84): deals with *extensional sets*: sets which are either *necessary* or *contingent*, without distinction.

extensional synthetic falsity (61): *dissimilarity* between *extensional meaning* and *reality*.

extensional synthetic truth (60): *similarity* between *extensional meaning* and *reality*.

extensional tautology (105): the extensional universe of discourse. Because *precursor extensional truth* is *set-membership*, the universe of discourse is always true. See also *extensional contradiction*.

extensional thought (56): *thought* with *extensions*. See also *pure thought*, *conceptual thought*, and *nominal thought*.

extensionally valid (60): an *extensional inference* from A to B is $A \subset B$ and is extensionally valid if and only if either (i) membership in A is universally membership in B: always if $x \in A$ then $x \in B$; or (ii) non-membership in B is universally non-membership in A: always if $x \notin B$ then $x \notin A$. Note that universality is *extensional necessity*, as opposed to singular possibility, which is *relational necessity*.

exotic relation (28, 131): an *ideal relation* between *possible worlds* as opposed to a *worldly relation* which is an ideal relation within a possible world.

extravagant multiplication (15): unwanted multiplication of the *existence* of *relations* that is not *endless multiplication*.

extrinsic necessity (10): *extrinsic possibility* that is singular. Examples are *demergence*, *emergence*, *inherence*, *causation*, and *implication*.

extrinsic possibility (44): two or more *relations* are extrinsically possible, relative to each other, if they may be terms of one relation: each is a coterm of the other in that relation. That is, if R and S are terms of Q, so that R is a coterm of S in Q and S is a coterm of R in Q, then R and S are extrinsically possible. A broader definition is: two relations are extrinsically possible if they may both exist in one possible world; this follows transitively from the first definition. See also *extrinsic necessity*.

extrinsic properties of relations (18): either *lower extrinsic properties* or *upper extrinsic properties*. The lower extrinsic properties of a relation are its *terms* and its upper extrinsic properties, or *uppers*, are the *relations* of which it is a term, as well as the other terms of these uppers. See also *coterms*, *intrinsic properties* and *extrinsic possibility*.

falsity (53): the absence of *truth*, of which there are the following kinds: *relational* and *extensional synthetic truth* are *similarity* between idea or proposition and reality; *relational analytic truth* is containment of predicate in subject, but not the result of a contradiction produced by its denial; *extensional analytic truth* is subset; *nominal truth* is correct linguistic usage; also *relational precursor truth* is *consistency*, *precursor extensional truth* is *set membership*, and *precursor nominal truth* is sense, so that *precursor nominal falsity* is nonsense.

fallacy of illicit major (227): an invalid *syllogism* in which the *major term* is *undistributed* in the *major premise* and *distributed* in the *conclusion*.

fallacy of illicit minor (227): an invalid *syllogism* in which the *minor term* is *undistributed* in the *minor premise* and *distributed* in the *conclusion*.

fallacy of undistributed middle (226): an invalid *syllogism* in which the *middle term* is *undistributed* in both *premises*.

fashion (198): fashionable people are following the dictates of their *oge* concerning the latest fashion. Leaders of fashion are *ego-dominant* and impose their ideas on their *oges*.

feeling (190): consciousness of *hekergies*; see also *thought*.

feeling of being watched (198): having this feeling when alone is explained by the fact that the *oge* is watching. The feeling is naturally particularly strong when performing guilty actions.

figure (226) of a *syllogism*: a number from 1 to 4, given by the arrangement of the *middle terms*. In numerical order the figures are MP, SM; PM, SM; MP, MS; and PM, MS.

first level subordinate terms (90): of a *relation* R are the *terms* of the terms of R.

ghosts (198): *oge-memories* of *oge-persons* causing *empirical* appearances in an *empirical world*.

goal (190): a goal, of an *agent*, is anything that will increase the *hekergy* of that agent.

gossip (199): is *oge* talking to oge.

Grand Structure (8): the noumenal, multilevel, structure of neutrons and protons, atomic nuclei, atoms, molecules, single-celled organisms, multicellular organisms, brains, consciousness, etc., with novel emergents at each level. It has contributions from all the theoretical sciences, and may be thought of as the ultimate unification of science.

greater than (91): a *relational natural number, n*, or M̂, is greater than another, *n* or N̂, symbolised *m>n*, if M̂≻N̂. Alternatively, *m>n* if there exist *representative instances* M and N of *m* and *n*, whose *term sets*, M and N, are such that N⊂M. It is not necessary for the relation of superset to exist between two term sets R and S in order to have a relation of greater than between their numbers: if R and S are such that R⊅S, R is equiadic with M and S is equiadic with N, and M⊃N, then *r>s*.

hekergy (37): if the number of possible *emergence-configurations* of a *set* of *terms* out of which a *relation* R *emerges* is *e*, the number of possible submergence-configurations of these terms is *s*, and *t = (s+e)*, then the hekergy of R is ln.(*t/e*). The numbers *e* and *s* are finite.

hypnotism (199): this is explained by the hypnotist putting the *ego* "to sleep" so that the *oge* may take control and make the individual behave according to the hypnotist's instructions; the hypnotist represents the *oge*. The fact that the hypnotist cannot make the individual behave immorally is due to the oge being the guardian of morals.

ideal (27): of the mind.

ideal actuality (134): the idea of *intrinsic necessity*.

ideal relations (26): *relations* in the mind, which are thereby *abstract ideas*.

idempotence (86): is the claim that connectives such as $A \cup A = A \cap A = A$ are true; it requires that *union, intersection*, and *set identity* may be monadic — and there are no *monadic relations*, except nominally; but idempotence of *intensions* such as Â≈Â is meaningful because one *instance* of Â may be *similar* to another instance.

identity (69): oneness, singularity; not to be confused with *similarity*, which requires plurality; see *identity error*.

identity error (69): the fallacy of inferring *identity* from *similarity*; to confuse identity and similarity, as if they were synonyms; identity is oneness, whereas similarity requires twoness, or greater, plurality.

illusion (105): *empirical falsity*.

image (53): a copy, representation, facsimile, or reproduction.

imagination (25): the part of the mind dealing with the *concrete*.

immediate inference (225): an inference which goes immediately from premise to conclusion, such as *conversion* and *obversion* in *traditional logic*.

imperceptible (239): a *perceptible*, for me, now, is anything that I am *conscious* of, now; and an imperceptible, for me now, is anything not a perceptible for me now. 'Me' and 'now' refer to whomever is thinking about this, and when. The *concept* of imperceptible is useful in describing *solipsism*.

implication theorem, the (79): $(\hat{P} \succ \hat{Q}) \Leftrightarrow (\{\mathcal{E}\hat{P}\} \subset \{\mathcal{E}\hat{Q}\})$.

impossibility (42): a zero *possibility*; see *modalities*.

improper subset and superset (72): if S is either a *subset* of T, or *identical* with T, this is symbolised by $S \subseteq T$, and its inverse by $T \supseteq S$; these are called improper subset and improper superset.

improper subintension and superintension (72): if $\hat{s}$ is either a *subintension* of $\hat{r}$ or is *similar* to $\hat{r}$, this is symbolised

275

$\hat{s} \divideontimes \hat{r}$, and its inverse by $\hat{r} \divideontimes \hat{s}$; these are called improper subintension and improper superintension.

incomplete disjunction (76): a *relational commonality*, or *disjunction*, whose *extensional meaning* is a *contingent set*, as opposed to a complete disjunction whose extensional meaning is a *necessary set*.

inherence (20): a *necessity* between a *relation* and all of its *properties*; the properties inhere the relation and the relation *exheres* the properties. The necessity is seen from the fact that the properties cannot exist without the relation.

instance-intensions: (49): *intensions* that define by *instances of relations*; see also *kind-intensions*.

instance of a relation (18, 19): an instance of a *kind* of a *relation* is determined by the *terms* and/or the *uppers* of the relation.

intension (3, 29, 48): the intension of an *intensional set* is the *commonality* of its plurality: all those *properties* possessed by all and only the members of the set. More technically, an intension is a relation, R, and one term, T, of that relation, such as R_T, such that the intension defines the set $\{\mathcal{E}R_T\}$. Kinds of intensions include *similarity-, dissimilarity-, kind-, instance-, demergence-, emergence-, inherence-,* and *boundary-intensions*.

intension theorem, the (78): for any *intension*, $\hat{P}$, $\bigwedge\{\mathcal{E}\hat{P}\} \wr \hat{P}$.

intensional logic (59): a portion of *relational logic* (logic that has *relational meaning*); the other portion being *mathematical logic*.

intensional set (29): a set which has an *intension*; also called a *necessary set* because its membership is necessitated by its intension and the *relation-every*. See also *extensional set*, and *contingent set*.

intensional set theorem, the (78): $(S=\{\mathcal{E}(\wedge S)\})\Leftrightarrow(S$ is an *intensional set*).

intersection (74): of two *sets*, S and T, symbolised by $S\cap T$, if it exists, is such that each *member* of $S\cap T$ is *identical* both with a member of S and with a member of T. If the intersection of S and T does not exist, S and T are said to be *disjoint*. (Note: the *null set*, which is commonly said to be the intersection of disjoint sets, is neither an *intensional set* nor a *contingent set*; it is only a *nominal set*.)

intrinsic necessity (134): *actuality*, or actual existence, as opposed to *intrinsic possibility* which is *mathematical* or *logical existence*; the former is *singular possibility*, the latter is *plural possibility*.

intrinsic possibility (45): an *intrinsic property* of all *relations*, also called *consistency* and *existence*. One of the three members of the *minim*. See also *extrinsic possibility* and *intrinsic* and *extrinsic necessity*.

intrinsic property (16): a *property* of a *relation*, as opposed to its *extrinsic properties*; the set of all the intrinsic properties of a relation *inhere* that relation.

intrinsic property set (45): the intrinsic property set of a *relation* R is the *intensional* set of every *intrinsic property* of R; it is usually symbolized by the same letter as R, as an italic capped small cap to the left of R: $\hat{R}R$; it is a *natural set*.

intuition (190): the ego may become aware of other parts of the *noumenal brain* and in doing so becomes conscious of some of their content. Two such parts are the *psychohelios* and the *oge*. Intuiting the former gives solutions to problems, and intuiting the latter gives moral convictions and revelations.

irrational (190): any assemblage of *noumenal ideas* that are arranged by *L.A.L.* (like-attract-like-and -repel-unlike), as opposed to the *rational*. A young *ego* is entirely irrational, while a mature ego may become partly rational through its desire to increase *hekergy*, in accordance with the *mind hekergy principle*. The *oge* is also mostly irrational.

join (42): a *relation* that has only the *minim* in its *intrinsic property set*.

kind-intensions: (49): *similarity-intensions* and *dissimilarity-intensions*; they define by *kind of relation*. All other intensions are *instance-intensions*: they define by *instances of relations*.

kind of a relation (16, 19): the *intrinsic properties* of a relation determine the kind of that relation, as opposed to an *instance* of a relation which is determined by its *extrinsic properties*.

L.A.L. (182): abbreviation for the principle that like *noumenal ideas* attract like and repel unlike noumenal ideas; with another factor, in that the *degree of similarity*, or likeness, is also a parameter.

level (24): If a *relation* R has level-n, where n is a number, then the *terms* of R are at level-$(n-1)$, the next lower level, and the *uppers* of R — the relations of which R is a term — are at level-$(n+1)$, the next higher level. See also *axiom level* and *top level*.

level-1 (24): the lowest *structure level*; also called the *axiom level*. The level on which all *relations* are *separators*.

linear separator (100): a *separator* that has unit length and is symmetric.

logical existence (17): *mathematical existence* or *possibility*.

logical possibility (42): the primitive property of *possibility* that is in the *minim* and is also *mathematical existence*.

logical proof: (232): consists of one or more given statements, called premises, a conclusion, and intermediate inferences establishing the truth of the conclusion, given the truth of the premises. The proof is set out formally in a series of lines, each consisting of three parts: a line number, for later reference, a logical statement, and a logical justification of that statement.

loop necessity (135): if a *possible world* has a *relation* R that has *necessary actuality*, symbolised by ⊡R, then the *loop possibility* of that world becomes loop necessity. Such a world is thereby an *actual* world.

loop possibility (32): *closed possibility*: every *whole* has closed possibility, in that its *top relation*, T, *downward necessitates*, by *demergence*, the *possibility* of all the *lower structures* of the whole (since a *relation* cannot *exist* without the existence of its *terms*), including its *axiom level*, which in turn *upward necessitates*, by *emergence* and *inherence*, the possibility of the rest of the whole. This also occurs with *separator loops* and *minor loops*.

loops (31): closed chains of *relations*, such as *separator loops* and *minor*, *middle*, and *major loops*.

lower extrinsic properties of a relation (18) are the *terms* of that relation. See also *uppers*.

love (191): a willingness, by the ego to give unconditionally to its beloved. This is only possible if the ego incorporates the beloved *mid-person* into the *ego*, so that loving it is an ego *hekergy* increase.

major loops (31): a *middle loop* that includes the *top relation* of a *possible world*. See also *middle loops* and *minor loops*.

major term (226): the *predicate* of the *conclusion* of a *categorical syllogism*. See also *minor term* and *middle term*.

major premise (226): the *premise* of a *syllogism* which contains the *major term*; it is always written first.

malice (199): provides an *illusion* of triumph over the *oge*, by diminishing the oge relative to the *ego*, as in practical jokes, bullying, and vandalism.

mapping (191): in the *noumenal mind* mapping is a copying of *noumenal ideas*, in whole or in part.

mathematical existence (17): *possibility*; within a given mathematical system, a particular mathematical entity may be possible, in which case it exists in that system, or it may be impossible, in which case it does not exist in that system. Also called *logical existence*. Possibility is one of the three *intrinsic properties* in the *minim*.

mathematical logic (57): logic that, with *intensional logic*, is *relational logic*; the logic used by mathematicians.

member of a set (3, 48): a **member**, defined by an *intension* R_T, is any *coterm* of T in R; R_T, with the *relation-every*, also defines the *intensional set* {*&*R<small>T</small>} of which a coterm, x, of T in R is a member: x ∈ {*&*R<small>T</small>}. See also *set-membership*.

middle loops (31): a middle loop consists of the *top relation*, T, of a *whole*, which *demerges minor loops* all the way down to, and including, the *axiom level*; and the axiom level *cascadingly emerges* all the minor loops up to, and including, T. See also *minor loops* and *major loops*.

middle term (226): the third *term* of a *syllogism*, after the *major* and *minor terms*; the term that occurs in each *premise* but not in the *conclusion*.

mid-body (183): an image of the noumenal body, causally midway between noumenal body and empirical body,

mid-memories (182): *noumenal ideas* that are causally midway between *noumenal memories* and *empirical memories*.

mid-object (181): a structure of *mid-sensations*, so called because it is causally midway between a *noumenal object* and an *empirical object*.

mid-sensation (181): a transient structure of neural switchings, so called because it is causally midway between a *noumenal property* and an *empirical sensation*.

mid-world (181): a structure of *mid-objects*, so called because it is causally midway between the *noumenal world* and the *empirical world*.

mind hekergy principle (180): the principle that the *noumenal mind* increases its *hekergy* whenever possible, and if this is not possible, it works to maintain its hekergy whenever possible, and if this is not possible, to minimise its loss of hekergy.

minim (16): the *set* of the three *intrinsic properties* possessed by every *relation* and only by relations: they are *possibility*, *simplicity*, and an *adicity*; the *commonality* of two *disparate relations* that have no *intrinsic properties* in common other than the fact of being *relations*; If Â, Ŝ, and P̂ represent the intrinsic properties of adicity, simplicity, and possibility then $\hat{M}ℵ(Â ⋎ Ŝ ⋎ P̂)$, where $\hat{M}$ is the symbol for the minim.

minor loops (31): a minor loop occurs with minimal *emergence*, *inherence*, and *demergence*: an *emergence-configuration*, C, emerges a set of properties, P̂, of a *relation*, R, which set inheres R, which R demerges its *terms*, which partially determine their configuration, C. See also *major loops* and *middle loops*.

minor premise (226): the *premise* of a *syllogism* which contains the *minor term* (the subject of the conclusion); it is always written after the *major premise*.

minor term (226) the *subject* of the *conclusion* of a *syllogism*. See also *major term* and *middle term*.

modalities (42): traditionally, *contingency*, *necessity*, and *impossibility*, defined respectively as plural, singular, and zero *possibility*.

modal logic (42): a branch of modern logic in which the *possible* is defined as anything that is true in at least one possible world, and the *necessary* is defined as anything that is as true in all possible worlds.

modal possibility (42): the possibility that may be quantified as plural, singular, and zero possibility, to give the three traditional *modalities*. See also *possibility*.

modal probability (42): the reciprocal of *modal possibility*; not to be confused with statistical probability — relative frequency — and the probability which is strength of belief in an explanation.

monadic relation (16): a *relation* with only one *term*; if any exist they can only exist in the axiom-level because their one term cannot be an *emergence-configuration*; so, like *separators*, they do not emerge but are *contingent* in level-1. They may also be *illusions* in ideal structures. Most probably, all monadic relations are nominal only, at best having meaning by analogy.

mood (226) of a *syllogism*: the conventional letter, A, E, I, or O, of the *major premise*, *minor premise,* and *conclusion*, in that order. Thus "SaM, and MeP, therefore SeP" has mood AEE.

moral (194): the characteristic attitude of the *oge*: the good (*hekergy* increase) of other mid-people, of society.

multiplication (92, 93): the multiplication of two *relational numbers*, m and n, symbolised $m \times n$ or mn, is the repeated *addition* of n instances of m: $m \times n = m_1 + m_2 + ... m_n$. Multiplication is thus a *compounded relation*.

naive realism (165): the error that the *empirical world* and the *noumenal world* are *identical*.

natural sets (29): a natural set of a relation, R, is a unified *plurality* of every *intrinsic property inhering* in R, or of every *term* of R, or of every *upper* of R, each plurality being unified into a *set* by the *relation-every*; these sets are the *property set*, the *term set*, and the *upper set*, respectively, of R.

necessary actuality (136): actuality, $\boxdot$, is *singular possibility* in the *minim*; when the property of *possibility* in the minim of a *relation* R changes from *plural possibility* to singular possibility, the existence of R changes from *mathematical existence* to *actual existence*. Necessary actuality, $\dot{\Box}\boxdot$, occurs when an actual relation *extrinsically necessitates actuality*.

necessary set (29): another name for an *intensional set*, because every intensional set has its membership necessitated by its *intension* and the *relation-every*.

necessity (42): *singular possibility*, as opposed to *plural possibility* which is *contingency*; also *actual existence*, as opposed to *mathematical existence*. A third usage, not used in this book, is the definition, in *modal logic*, of necessity as that which is true in every possible world. See also *possibility*.

negation theorem, the (80): $\{\mathcal{E}\hat{P}\}'=\{\mathcal{E}\hat{P}'\}$, $\hat{P}\,\mathcal{R}\,\hat{P}''$, and $\{\mathcal{E}\hat{P}'\}'=\{\mathcal{E}\hat{P}\}$.

nominal analyticity (63): is consistency, in the sense that it is not self-contradictory or does not lead to a contradiction.

nominal arithmetic (99): the arithmetic of *nominal numbers*.

nominal definition (10): synonymy.

nominal falsity (105): incorrect statement of meaning, as in ignorance, error, deceit, or nonsense.

nominal function (112): an *extensional function* whose domain, range, or both, are *nominal sets*.

nominal logic (10, 63): *truth-functional logic* (see Appendix B for details). Criticisms of this logic are given in Chapter 3.

nominal meaning (10, 41): puts *relational meaning* or *extensional meaning* into words or symbols, but it may be only meaning by verbal analogy to either relational or extensional meaning, and have no reference; it occurs only in language; it allows paradox and contradiction. See also *relational meaning* and *extensional meaning*.

nominal necessity (107): the *necessity* of *truth-functional validity*, or tautology, as opposed to *relational necessity* which is *singular possibility*, and to *extensional necessity* which is universality.

nominal number (99): the name or symbol of any *relational* or *extensional natural number* for which a name or symbol exists, or any number defined nominally.

nominal relations (16, 28): names of *relations*; *purely nominal relations* are those that exist only in language; they can be named and described, but such language has no reference.

nominal set (29): a name or description of a *set* which may have reference to an *intensional set* or an *extensional set* or have no reference at all.

nominal set theory (88): deals with *nominal sets*: sets which are names of sets which are *intensional sets, extensional sets* or *nominal sets.*

nominal thought (56): thought with words or symbols alone, silent speech. Compare with *pure thought*, which is thought with abstract ideas alone; and *conceptual thought*, which is thought with abstract ideas bonded to words or symbols. Nominal thought is algorithmic.

nominal truth (63, 105): correct (i.e. established) usage of language, correct statement of meaning — *relational, extensional, nominal, ideal, noumenal* or *empirical* meaning.

nominalism (65): the doctrine that no *abstract ideas* exist.

nominally valid (63): A nominal inference of one statement, q, from another, p, is nominally valid if and only if the *truth function* p→q is *tautologous*, meaning always *nominally true*.

noumenal (26): that which exists and is underlying, theoretical, non-empirical.

noumenal actuality (135): the cause of *empirical actuality*; it is a state in the *minim* of *relations* in the *noumenal world*, in which the *intrinsic property* of *possibility* is *singular possibility*, or *necessity*.

noumenal brain (181): the brain in the noumenal body, not to be confused with the empirical brain

noumenal ego (183): a noumenal subject of consciousness; initially a *structure* of *mid-memories* of the *mid-body*, which later includes *beliefs*.

noumenal existence (28): *actual existence* in the *noumenal world*: *singular possibility*, as opposed to *mathematical possibility*.

noumenal-God (202): the *noumenal world*; one of the meanings of the word God.

noumenal idea (181): a structure of *atomic ideas*, which are *On's* and *Off's* in a *noumenal brain*, structures of these, structures of structures, and so on.

noumenal memories (182): *mid-memories.*

noumenal mind (171, 180): an *emergent* out of *noumenal brain.*

noumenal motor ideas (185): *noumenal ideas* that go to the noumenal efferent nerves in order to move noumenal muscles.

noumenal perception (191): a causal process producing noumenal images of noumenal objects in the *noumenal brain.* These images are *mid-objects* and produce *empirical objects.*

noumenal reality (54): all that *exists* in the *noumenal world.*

noumenal relations (26): *relations* underlying the *empirical world*; that is, relations in the *noumenal world*; relations that exist but are neither *empirical* nor *ideal.*

noumenal sensation (181): a transient structure of neural switchings, also called a *mid-sensation* because it is causally midway between a noumenal property and an empirical sensation.

noumenal world (7, 11, 27): the totality of *noumenal relations*, including the *Grand Structure.*

novel property (21): an *emergent property* of a *relation*, not possessed by any lower *level* relations: this property is defined as *dissimilarity* between the emergent property and all lower level properties.

null set (30, 74): a *set* which has no *extension*. It does not exist except nominally because it produces contradictions such as 'the intersection of non-intersecting sets'; it is also a *set relation* with no *terms*, hence not a set; and if it were an *intensional set* it would require an infinite *intension* to define it.

number one (90): relationally, simplicity; extensionally, the set of all one-membered sets; nominally, the name of each of these.

obversion (225): an operation on a *categorical proposition* that both negates the premise (by changing the predicate to its complement) and changes the quality (affirmative or negative) of the proposition. See also *conversion*.

oge (194): an anti-ego; a *noumenal agent* opposed to the *noumenal ego* and representing society and *morality*.

oge-dominant (196): a person whose *oge* dominates their ego; an oge-dominant person is more moral than selfish. See also *ego-dominant*.

oge-God (201): the *oge*; one of the meanings of the word God.

On's and off's (181): possible switchings between two noumenal neural cells; an *off* is a relation dissimilar to an *on*. They are atomic *noumenal ideas*.

one-to-one correspondence (5): two *sets*, P and Q, are in one-to-one correspondence if, for every *member* of P there corresponds one, and only one, member of Q, and *vice versa*. P and Q are then said to have the same *extensional number* of members, this number being the set of all sets that are in one-to-one correspondence with P.

ordered set (4): symbolized with angle brackets, the ordered pair ⟨x,y⟩ means that in every pair ⟨x,y⟩, x comes first and y comes second; an ordered pair is an *extensional relation*. An ordered set may have any number of terms, all ordered.

parts of a whole (26): if R is the *top relation* of a *whole* then the *subordinate terms* of R which are themselves top relations of wholes define the *parts* of the whole.

pattern diagrams (82): *relational meanings* may be represented graphically, by pattern diagrams, in a manner complementary to the graphical representation of *extensional meanings* by Venn diagrams.

perceptible (239): (noun) a perceptible, for me, now, is anything that I am *conscious* of, now; and an *imperceptible*, for me now, is anything not a perceptible for me now. 'Me' and 'now' refer to whomever is thinking about this, and when. It is a concept useful in understanding *solipsism*.

pleasure and pain (191): pleasure is *consciousness* of *hekergy* increase, pain is consciousness of hekergy decrease. These may be physical, due to hekergy changes in the *noumenal body* which are *mapped* into the *empirical body*, or mental, due to hekergy changes in the *noumenal mind* which come into consciousness.

plural possibility (7): contingency; one of the three traditional modalities, the others being *singular possibility* or *necessity*, and *zero possibility* or *impossibility*.

politics (199): *oge-dominant* people can be expected to be radical in politics: on behalf of their *oge* they favour the good of society, and particularly of the under-privileged, at the tax-payers' expense. *Ego-dominant* people can be expected to be reactionary in politics: they favour maximum opportunity for *selfish* gain.

polyadic predicate: (104) synonym for a *relation*. A description that occurs in modern logic: some logicians suppose that grammatically a subject may have one predicate or many predicates. This gives rise to what are called the monadic predicate calculus and the polyadic predicate calculus, two developments of truth-functional logic. See Appendix B.

population (132): of a *level* is the number of *instances* of *relations* in that level. See also *variety* of a level.

possibility (16, 42): There are three usages of the word possibility. One is the primitive property that is in the *minim* and is variously called *consistency* and *existence*, as well as possibility. The second is called *modal possibility*, it may be quantified as plural, singular, or zero, to give the three traditional modalities of *contingency*, *necessity*, and *impossibility*. These two meanings are combined in that the possibility in the minim may be either plural or singular; if it is plural it is *logical existence* and if it is singular it is *actual existence*. The third meaning of possibility is the circular definition of possibility as true in at least one possible world; this latter belongs to modern *modal logic*. See also *necessity*.

possible noumenal world (27), or **possible world** for short: an *ideal* relational *structure*; a *whole* whose *top relation* cannot be a *term* of another *relation* — the top relation has no *uppers* — because a single relation cannot form an *emergence-configuration*. A possible world is a candidate for a true description of the *noumenal world*.

precursor extensional truth and falsity (61): precursor extensional truth is *set-membership*, and precursor extensional falsity is non-membership; symbolised by $\in$ and $\notin$, or by $\wr$ and $\sim$.

precursor nominal truth and falsity (63): precursor nominal truth is sense, as opposed to precursor nominal falsity which is nonsense; symbolised by $\wr$ and $\sim$.

precursor relational truth and falsity (55): A necessary condition for both *relational analytic truth* and *relational synthetic truth* is *consistency*, which is one of the properties of the *minim* — also known as *possibility* and as *logical* or *mathematical existence*. This is precursor relational truth, symbolised by $\Diamond$ or by $\wr$, and its absence is precursor relational falsity, symbolised by $\sim\Diamond$ or by $\sim$.

prejudice (192): a belief that is strong enough to select evidence in its favour, and reject evidence against it, by *L.A.L.* (like-attracts-like-and-repels-unlike).

primitive magic (199): is various ways of dealing with the *oge*, all of which are displaced on to the *empirical world*. Sympathetic magic, such as with pins in a voodoo doll, and name magic — confusion between name and named — are misguided attempts to control things or people in the empirical world by manipulating the oge.

principle of duality (244): in mathematics it is the principle that if any axioms imply their own *duals* then any theorems implied by these axioms also imply their own duals.

probability (44): one of: *modal probability*, the reciprocal of *possibility*; statistical probability (relative frequency); or the probability which is strength of *belief* in an explanation.

problem of induction (107, 292): the problem of how to justify reasoning from some to all; the major epistemological difference between *relational necessity* (singular *possibility*) and *extensional necessity* (universality), in that extensional necessity is limited by the problem of induction while relational necessity is not.

property (16): in this book refers to *intrinsic properties* of *relations*. There are also upper and lower extrinsic properties of relations, which are their terms and their uppers. All are primitive, undefined.

property set (45): the *set* of every *property* of a *relation*.

proposition (53): the meaning of an abstract statement; a *structure* of *abstract ideas*, which are *ideal relations*.

process (38): a sequence of transitive *causations*.

psychohelios (203): a third *agent* in the *noumenal mind*; a *structure* of *noumenal ideas*, at the periphery of the *ego*, arranged with maximum *hekergy*; a part of the noumenal mind that is wholly *rational*.

pure thought (55): *abstract thought* without the aid of language; that is, manipulation of *abstract ideas* — *ideal relations* — rather than *concepts*. Compare with *conceptual thought*, which is thought with abstract ideas bonded to words or symbols; *extensional thought* which is thought with (small) *extensions*, and *nominal thought*, which is thought with words or symbols alone.

purely nominal relations (16): *relations* that can be named and/or described, but the names and descriptions have no reference. Also called *exclusively nominal relations*.

quality of a possible world (37): the total amount of emergent *hekergy* in a *possible world*, divided by the number of *separators* (*axiom relations*) in its *axiom level*. It is proportional to the *level* of the *top relation* of the world.

quasi-separators (132): separators such as *atomic squares*, which can separate atomic squares, but they are not *axiom separators* because chains of them do not have to form *closed loops.*

Q.Q. (70): abbreviation for the principle that qualitative difference entails quantitative difference.

rational (192): *noumenal ideas* that are arranged with maximum *hekergy* are rationally arranged, as opposed to ideas that are arranged by *L.A.L.*, which are *irrationally* arranged. Maximum hekergy occurs when ln.t/e is a maximum, meaning that $e=1$; and $e=1$ is a singular possibility, which is a necessity; and necessity in thought is rational; also, maximum hekergy is maximum *value.*

reality (53): either *empirical reality*, which is everything empirical which is not illusory, or else *noumenal reality*, which is everything in the *noumenal world.*

recognition (192): If a *mid-object* attracts a *mid-memory* to itself because they are *similar* this similarity is recognition: mid-recognition and empirical recognition.

relation (16): A primitive concept, characterised by having at least the three intrinsic properties of an *adicity*, *simplicity*, and *possibility*, which together constitute the *minim.*

relation-any (46): the *relational function* which has *sets* as its arguments and *members* as its values; its inverse is the *relation-every.*

relational analytic truth (54, 60, 63): *superintension*; the subject contains the predicate: $\hat{s} \succ \hat{P}$. See also *extensional analytic truth*.

relational complement (75) of ŝ is ≈ŝ, also symbolised by ŝ′, which reads as non-ŝ. If $\hat{v}$ is the universe of discourse of intrinsic properties then $\hat{v}$-{Ɛℵŝ}ℵ{Ɛℵŝ′}ℵ{Ɛ≈S}. See also *extensional complement*.

relational conjunction (73): *coupling* of *intensional sets*.

relational connectives (71): *relations* between *sets* of *intrinsic properties* which are connectives defined by means of *similarity*; see also *extensional connectives*, which are defined by means of *identity*.

relational definition (10): is definition of words that have *relational meaning*.

relational disjunction (74): *commonality* of *intrinsic property sets*.

relational equivalence (71): *similarity* of *intrinsic property sets*.

relation-every (46): the *relational function* which has *members* as its arguments and *sets* as its values; its inverse is the *relation-any*.
relational function (110): any *relation* whose *intrinsic property set* is a *superintension* of the property of *necessity*, of *singular possibility*.

relational implication (73): *superintension*.

relational logic (10): *intensional logic* together with *mathematical logic*.

relational meaning (10, 40): a symbol, name, or description has relational meaning if its meaning is a *relation*, or one or more *properties* of a relation. See also *extensional meaning* and *nominal meaning*.

relational meaning of an extensional number *n* (98): the *intension* of that number; that is, the intension of the *set* of all *n*-membered *extensional sets*, which is the intension of the set of all the extensional sets that are in *one-to-one correspondence* with an *n*-membered set.

relational necessity (106): singular possibility, as opposed to *extensional necessity* which is universality, and *nominal necessity* which is tautology in truth-functional logic, and 'that which is true in every possible world' in truth-functional modal logic.

relational number (90): an *adicity* of a *relation*.

relational number of a set *S* (98): is the *relational number* of any *term set* that is in *one-to-one correspondence* with *S*.

relational number one (90): *simplicity*, which is essentially unity. Every *relation* possesses the *minim*, which includes the properties of simplicity and *adicity*, so is a *representative instance* both of the number one and of an adicity, as is shown by it being one relation and by the number of its term set.

relational number of a contingent set (98): the *relational number* of any *intensional set* with which the contingent set is in *one-to-one correspondence*.

relational number of an intensional set (91): the *adicity* of any *relation* whose *term set* is in *one-to-one correspondence* with that intensional set.

relational precursor truth and falsity (55): A necessary condition for both *relational analytic truth* and *relational synthetic truth* is *consistency*, which is one of the properties of the *minim* — also known as *possibility* and as *logical* or *mathematical existence*. This is precursor relational truth, symbolised by ◊ or by ₹, and its absence is precursor relational falsity, symbolised by ~◊ or by ~.

relational set theory (69): is the theory of *intensions* and their connectives, relationally defined *extensions* and their connectives, and of the *relations* between these two kinds of connectives.

relational synthetic falsity (53): if a *relation, structure*, or *whole*, is an *image*, copy, representation, or reproduction of another, and they are *dissimilar*, then their dissimilarity is called the relational synthetic falsity, of the copy, relative to the other, or original. Also called *dissimilarity falsity*.

relational synthetic truth (53, 55): If a *relation, whole*, or *structure* is an *image* copy, representation, or reproduction of another, and they are *similar*, then their similarity is called the relational truth of the copy, relative to the other, or original. Also called similarity truth. If they are not similar then the copy is *relationally false*, or *dissimilarity false*, relative to the original. See also *relational analytic truth*.

relational universe of discourse (76): There are three possible relational universes of discourse. The largest, $U_{\dot{P}}$, is the set of all *intrinsic properties* of *relations*; next is U_R, the set of all *kinds* of *relations* plus all *instances* of each kind; and third is $U_{\dot{P} \succ \dot{M}}$, the set of all kinds of relations. Clearly, $U_{\dot{P}} \supset U_R \supset U_{\dot{P} \succ \dot{M}}$. The last, $U_{\dot{P} \succ \dot{M}}$, is the one of most interest in this book.

296

relational validity (57) is *necessary* transmission of *relational truth* or *falsity*; the inference from Â to B̂ is relationally valid if and only if the relational truth of Â *relationally necessitates* the *truth* of B̂: ⟨Â ⇒⟨B̂, or the relational falsity of B̂ relationally necessitates the falsity of Â: ~Â ⇒ ~B̂.

R⇒E⇒N (41): abbreviation for the fact that *relational meaning* entails *extensional meaning* which entails *nominal meaning*.

representative instance (89): if a *relation* R has a *property set* R̂ then R is a representative instance of a property set ŝ if R̂ is a member of {ε(⋇ŝ)}, or R̂⋇ŝ. For example, a *relational conjunction* is a representative instance of each of its conjuncts.

rites of passage (199): baptisms, coming of age, graduations, weddings, and funerals are all, as rites of passage, necessarily public because they are social and so involve the *oge*.

S. B. S. (179): abbreviation for 'skull beyond sky'.

secondary qualities (175): empirical images of qualities that are manufactured in the *noumenal brain* of the perceiver as a result of receiving signals from the sense organs, via the afferent nerves.

selection (39): an operation comparable to the *comparison* that causes higher-level *similarities* to *emerge*.

self (193): an *emergent noumenal idea* that unifies the collection of *mid-memories* of the *noumenal body* into a *whole*, a whole greater than the sum of its parts, a whole which is the *noumenal ego*; the *top relation* of the noumenal ego.

self-sacrifice (200): is mis-named: it is sacrifice of the *ego*, by the *oge*, for the good of society.

self-similarity (14): a *relation* that *multiplies endlessly*.

selfishness (193): the *noumenal ego* needs to increase its own *hekergy*, in accordance with the *mind hekergy principle*; this need is the basic attitude of the ego, and is called selfishness.

separator (24, 30): a relation that is equally *relation* and *term* in the lowest level, the *axiom level*, or *level-1*. Separators are not *emergent* and do not have *hekergy* because they do not emerge from *emergent configurations*. See also *quasi-separator*.

separator loops (31): *axiom separators* form chains and these have to form closed loops, called separator loops.

set (3, 35): a *number* of entities. A set may be *natural, intensional* (or *necessary*), *extensional, contingent*, or *nominal*. See also *set relation*.

set equality (71): a common misnomer for *set identity*.

set difference (75): The set difference of two *intersecting intensional sets*, *S* and *T*, symbolised *S−T*, if it exists, is the set consisting of those *members* of *S* which are not identical with any member of *T*.

set identity (71): Two extensional sets, *S* and *T*, are identical, symbolised by *S=T*, if each member of *S* is *identical* with a member of *T*, and *vice versa:* $(S=T) \Leftrightarrow (\&x)[(x \in S) \Leftrightarrow (x \in T)]$.

set-membership (3, 48, 61): the *demergence* of a *term* of a *set relation*. It is the relation between a set and a member of that set; it is the *extensional meaning* of *analytic truth*; it is symbolised by ∈.

set relation (48): the compounded join of every member defined by an intension; as such it is a relation, called a set relation. It is symbolised by braces: {}. Thus an intensional set defined by an intension Rᴛ is {ƐRᴛ}, where Ɛ is the symbol for 'every'.

similarity (34): a primitive *relation*; it would multiply endlessly were it not restricted (i) by not having any *uppers* and (ii) by having only locally separated *terms*. It also emerges in higher-level structures such as minds or computers, as a result of comparisons. Symbol: ℛ. See also *dissimilarity*.

similarity-intension (49): a similarity and one of its terms R, as in ℛR, which defines a similarity-set: {ƐℛR} or, more fully, {ƐXℛR}, where X is a variable. This set is the set of every relation that is similar to R. Similarity-intensions are the basis of classification.

similarity set (49, 75): the *intensional set* consisting of every *relation* having a *property set* similar to a given property set, as in {Ɛℛŝ}. See also *dissimilarity set*.

similarity truth (53): If a *relation*, *whole*, or *structure* is a copy, *image*, representation, or reproduction of another, and they are similar, then their similarity is called the similarity truth, of the copy, relative to the other, or original. Also called *relational truth* and *correspondence truth*. If they are not similar then the copy is *relationally false*, or *dissimilarity false*, relative to the original.

simplest possible world (131): a *possible world* that has a level-2 *top relation* and consists of four *linear separators*, each of which separates two *right angle separators*, each of which separates two of the linear separators; and between each pair of linear separator and angle separator is a *dissimilarity*; out of this *emerges* the top relation, called a square, which defines a *whole* having the *emergent* property of an area.

simplicity (16): one of the *properties* in the *minim*, possessed by all *relations*, which requires that they each be one relation, indivisible; the origin of the *relational number one*.

singular possibility (7, 42, 43): necessity; one of the three traditional modalities, the others being *plural possibility* or *contingency* and *zero possibility*, or *impossibility*. *Possibility* is one of the three primitive intrinsic properties in the *minim*: if it is plural it is *mathematical existence*, if it is singular it is *actual existence*.

sleep walking (200): when someone is sleepwalking there is no puzzle as to how they can control their body while genuinely asleep: they do not control their body, the *oge* does. The same happens with sleep-driving.

solipsism (171, 239): the doctrine that I alone exist, which arises from the claim that no *imperceptibles* exist. See Appendix C.

speculation in science (108): theoretical science is speculative regarding *synthetic truth* while empirical science is speculative regarding analytic truth.

square of opposition (225): a graphic mnemonic which shows the truth or falsity of all categorical propositions — A, E, I, or O — having the same subject and predicate.

stationary principle (167): a principle that requires that a rate of change be a maximum, a minimum or a point of inflexion.

structure (8, 25): a structure consists of *levels* of *relations*, those relations of one level being the *terms* of those at the next higher level. Its top level is a single relation, called its *top relation*. See also *whole* and *possible world*.

subintension (72): A *property set*, $\hat{s}$, is a subintension of another property set, $\hat{r}$, symbolised $\hat{s} \prec \hat{r}$, if each *member* of $\hat{s}$ is *similar* to a member of $\hat{r}$, but not *vice versa*. The inverse of subintension is *superintension*, symbolised by $\succ$. $\hat{s} \prec \hat{r}$ means that for all $\hat{x}$, $(\hat{x} \in \hat{s}) \Rightarrow (\hat{x} \in \hat{r})$ but $(\hat{x} \in \hat{r}) \not\Rightarrow (\hat{x} \in \hat{s})$. If $\hat{s}$ is either a subintension of $\hat{r}$ or is similar to $\hat{r}$, this is symbolised $\hat{s} \preccurlyeq \hat{r}$, and its inverse by $\hat{r} \succcurlyeq \hat{s}$; these are called *improper subintension* and *improper superintension*.

submergence (14): the going out of *existence* of a relation due to the *configuration* of its *terms* changing from an *emergence-configuration* to a *submergence-configuration*.

submergence-configuration (20): a configuration of the terms of a relation which necessitates the submergence, or else continued non-existence, of that relation. See also *emergence-configuration*.

subordinate adicity (92): if a *relation* R has *adicity r*, then the first level subordinate adicity of R is the sum of the adicities of each of the *r terms* of R: the total number of the terms of the terms of R. The second level subordinate adicity of R is the total number of the terms of the terms of the terms of R, and so on.

subordinate levels of a relation (90): if the emergent level of a relation R is *n*, #R=*n*, then the emergent level of the terms of R, *n-1*, is the first subordinate level of R, the emergent level of the terms of the terms of R, *n-2*, is the second subordinate level of R, and so on.

Relations

subordinate terms (26): of a *relation* R are the *terms* of R, the terms of their terms, and so on down to the lowest *level*. The first level subordinate terms of R are the terms of R.

subset (72): an *intensional set*, S, is a subset of another intensional set, T, symbolised $S \subset T$, if each *member* of S is *identical* with a member of T, but not *vice versa*. The inverse of subset is *superset*, symbolised by $\supset$. $S \subset T$ means that for all x, $(x \in S) \Rightarrow (x \in T)$ but $(x \in T) \not\Rightarrow (x \in S)$. If S is either a subset of T, or identical with T, this is symbolised by $S \subseteq T$, and its inverse by $T \supseteq S$; these are called *improper subset* and *improper superset*.

sum (92): (a special case): a *sequence* of *joins* between the *terms* of an n-adic relation, R, each *term* being a unity and so a *representative instance* of the *relational number one*, such that $1_1 + 1_2 + ... 1_n$ is their sum. See also *addition*.

superintension (72): the inverse of *subintension*, symbolised $\hat{S} \prec \hat{T}$.

superset (72): the inverse of *subset*, symbolised $S \subset T$.

suicide (200): is the killing of the *ego*, by the *oge*, for the good of society.

supra-rational (203): Platonic wisdom; a state of consciousness that is supra-linguistic, timeless, and of maximum possible value.

syllogism (226): an *argument* in *traditional logic*.

synthetic truth (52): *similarity truth*; also called *relational truth* and *correspondence truth*.

temporal separator (101): a *separator* that has unit duration, and is asymmetric.

terms (224): the *lower extrinsic properties* of a *relation*. One of the characteristics of a relation is that it has terms; all relations, and only relations, have terms. The terms of a relation *demerge* from their relation. The word term is also used in *traditional logic* to refer to either a subject or a predicate in a *categorical proposition*.

term set (19): the term set of a *relation* R is the *natural set*, and so also the *intensional set*, consisting of every *term* of R; it is symbolized by the same letter, small cap, italic: R or RR.

theoretical (7): non-empirical, underlying, imperceptible.

Theorem 4.1 (78): the intension theorem: for any intension, $\hat{P}$, $\wedge\{\mathcal{E}\hat{P}\}\,\mathcal{U}\,\hat{P}$.

Theorem 4.2 (78): the extension theorem: for any set S, if $\wedge S$ exists then $S\subseteq\{\mathcal{E}\wedge S)\}$.

Theorem 4.3 (78): the intensional set theorem: an extension S is an intensional set if and only if $S=\{\mathcal{E}(\wedge S)\}$

Theorem 4.4 (79): the contingent set theorem: an extension S is a contingent set if and only if either $\wedge S$ does not exist, or else $\wedge S$ exists and $S\subset\{\mathcal{E}(\wedge S)\}$.

Theorem 4.5 (79): the equivalence theorem: $(\hat{P}\,\mathcal{U}\,\hat{Q})\Leftrightarrow(\{\mathcal{E}\hat{P}\}=\{\mathcal{E}\hat{Q}\})$. Corollary: $(\hat{P}\approx\hat{Q})\Leftrightarrow(\{\mathcal{E}\hat{P}\}\neq\{\mathcal{E}\hat{Q}\})$.

Theorem 4.6 (79): the implication theorem: $(\hat{P}\succ\hat{Q})\Leftrightarrow(\{\mathcal{E}\hat{P}\}\subset\{\mathcal{E}\hat{Q}\})$.

Theorem 4.7 (79): the conjunction theorem:
$\{\mathcal{E}\hat{A}\curlyvee\hat{B}\}=\{\mathcal{E}\hat{A}\}\cap\{\mathcal{E}\hat{B}\}$.

Theorem 4.8 (79): for all intensional sets *A* and *B*,
$\wedge(A\cup B)\aleph[(\wedge A)\curlywedge(\wedge B)]$.

Theorem 4.9 (80): the disjunction theorem:
$\{\mathcal{E}(\hat{P}\curlywedge\hat{Q})\}\supseteq\{\mathcal{E}\hat{P}\}\cup\{\mathcal{E}\hat{Q}\}$.

Theorem 4.10 (80): the disjunctive addition theorem:
$\{\mathcal{E}\hat{P}\}\subset\{\mathcal{E}\hat{P}\}\cup\{\mathcal{E}\hat{Q}\}$ if and only if the disjunction is complete,
which is true if and only if $\{\mathcal{E}\hat{P}\}\cup\{\mathcal{E}\hat{Q}\}$ is an intensional set.
And the disjunction is incomplete if and only if $\{\mathcal{E}\hat{P}\}\cup\{\mathcal{E}\hat{Q}\}$
is a contingent set.

Theorem 4.11 (80): **the negation theorem**: $\{\mathcal{E}\hat{P}\}'=\{\mathcal{E}\hat{P}'\}$,
$\{\mathcal{E}\hat{P}'\}'=\{\mathcal{E}\hat{P}\}$, and $\hat{P}\aleph\hat{P}''$.

Theorems 4.12 and 4.13 (81): the distribution theorems:
$\hat{P}\curlywedge(\hat{Q}\curlyvee\hat{R})\aleph[(\hat{P}\curlywedge\hat{Q})\curlyvee(\hat{P}\curlywedge\hat{R})]$ and $\hat{P}\curlyvee(\hat{Q}\curlywedge\hat{R})\aleph[(\hat{P}\curlyvee\hat{Q})\curlywedge(\hat{P}\curlyvee\hat{R})]$.
;
Theorem 9.6. (152): $\boxdot$**G**. The best of all possible worlds is
actual.

thought (25, 193): manipulation of *abstract ideas* according
to their meanings, as opposed to feeling, which is their
manipulation according to their *hekergies* and which is
experienced in consciousness as subjective *value*. See also
imagination, and *abstract*, *pure*, *conceptual*, and *nominal*
thought.

top down necessity (21): *downward necessity*; *demergence*;
the *necessity* which makes the *existence* of a *relation* require
the existence of its *terms*.

top level (24): the highest *level* of a *structure* or a *possible world*; it consists of a single *relation*.

top relation (25): the highest level *relation* of a *structure* or of a *whole*; the relation which unites the parts of a structure, of a whole, or of a possible world. The difference between a structure and a whole is that the top relation of a whole has a novel property, not existent in any lower level of the whole, and which makes the whole greater than the sum of its parts. A possible world is a whole whose top relation has no *uppers*.

traditional logic (224): the logic which originated with Aristotle and was systematized by medieval philosophers; it is based on *categorical propositions*, of which there are four kinds: 'All S are P', 'No S are P', 'Some S are P', and 'Some S are not P', where 'S' stands for subject and 'P' for predicate; S and P are called *terms* and each refers to a class of entities, so that sometimes this logic is called a logic of classes. See Appendix A.

truth (53): *relational synthetic truth* is *similarity* between idea or proposition and reality; *relational analytic truth* is containment of predicate in subject, $\hat{s} \succ \hat{p}$, but not the result of a contradiction produced by its denial. *Extensional analytic truth* is subset; *extensional synthetic truth* is relational synthetic truth. *Nominal truth* is correct linguistic usage. Also *precursor relational truth* is *consistency, precursor extensional truth* is *set membership* and *precursor nominal truth* is sense, as opposed to *precursor nominal falsity*, which is nonsense.

truth-functional logic (63, 231) a pseudo-logic in which everything is defined in terms of *truth* and *falsity*. See Appendix B.

truth-functional validity (63): tautology.

truth-functions (231): *relations* which are sentential *connectives* in *truth-functional logic*.

underlying (7): theoretical, non-empirical, noumenal.

undistributed (224): a *term* is *distributed* in a *categorical proposition* if in that proposition it refers to the entire class named, and otherwise it is undistributed.

union (73): the union of two *intensional sets*, *S* and *T*, symbolised by $S \cup T$, is such that each *member* of $S \cup T$ is *identical* either with a member of *S* or with a member of *T*, or both.

unit magnitude (99): *axiom relations* (*level-1 separators*) are *compoundable relations* and as such have *intrinsic properties* which sum. These properties are unit magnitudes, unit measures.

upper extrinsic properties of relations (18): the upper extrinsic properties, or *uppers*, of a *relation* R are all the relations of which R is a *term*, plus the terms (other than R) of these relations, which are the *coterms* of R in these relations.

uppers (18); the *upper extrinsic properties* of a *relation*.

upper set (19): the *set* of all *upper extrinsic properties*, or *uppers,* of a *relation*. All relations of all *possible worlds*, except *top relations* of these worlds, have such a set. Also called an upper extrinsic set.

upward necessity (21): is *emergence* of *intrinsic properties*; if the *terms* of a *relation* R exist in one *level*, and are suitably *configured*, then the properties of R necessarily emerge in the next higher level, and from them inhere R. Also called *bottom up necessity*.

upper set (19): short for *upper extrinsic set*.

valid (60, 63): an *argument* is *relationally valid* if the truth of the premises *necessitates* the truth of the conclusion, or the falsity of the conclusion necessitates the falsity of at least one of the premises. An argument from *set A* to set *B* is $A \subset B$ and is *extensionally valid* if and only if either membership in *A* is universally membership in *B* or non-membership in *B* is universally non-membership in *A*. Note that *relational necessity* is singular possibility, as opposed to *extensional necessity*, which is universality. In *truth-functional logic* an argument is *nominally valid* if it is a *tautology*.

valid argument forms (57): are forms of logical argument that are generally accepted as valid, and which may be quoted in a formal argument; relationally they contain necessity, ⇒, ⇑ or ⇓, which justifies their validity; extensionally, validity is subset; and nominally it is tautology.

valid equivalences (57): relationally are symmetric argument forms; they are expressions containing the binecessity, ⇔, either side of which may be substituted for the other in a logical argument. Similarities, such as $\hat{s} \,⅋\, \hat{P}$, are also a basis for equivalence. Extensional equivalences are so-called set equalities, better called set identities; nominal equivalences are truth-functional equivalences: two statements are equivalent if they have the same truth value, and are otherwise not equivalent.

variety (132): of a *level* is the number of *kinds* of *relations* in that level and in all lower levels.

value (193): value is *hekergy*. Absolute values are *noumenal* hekergies, relative values are subjectively distorted *consciousnesses* of these. Traditionally these values were called *truth*, beauty and the good.

whole (26): A whole is defined by an emergent *top relation*, T, which possesses at least one *novel property* — a property that does not appear on any level lower than that of T. It is a special case of a *structure*, and a *possible world* is a special case of a whole.

worldly relation (28, 131): an *ideal relation* in a *possible world*, as opposed to an *exotic relation* which is an ideal relation between possible worlds.

zero (93): a nominal number that has no relational or extensional meaning.

zero possibility (7): impossibility; one of the three traditional *modalities*, the others being *singular possibility* or *necessity*, and *plural possibility* or *contingency*.

Index.

315

Draft 12.1, 2019/8/01.

www.ingramcontent.com/pod-product-compliance
Lightning Source LLC
Chambersburg PA
CBHW060323200326
41519CB00011BA/1822